Understanding RISK

Informing Decisions in a Democratic Society

Paul C. Stern and Harvey V. Fineberg, editors

Committee on Risk Characterization

Commission on Behavioral and Social Sciences and Education

National Research Council

NATIONAL ACADEMY PRESS
Washington, D.C.   1996

NATIONAL ACADEMY PRESS • 2101 Constitution Ave., N.W. • Washington, D.C. 20418

NOTICE: The project that is the subject of this report was approved by the Governing Board of the National Research Council, whose members are drawn from the councils of the National Academy of Sciences, the National Academy of Engineering, and the Institute of Medicine. The members of the committee responsible for the report were chosen for their special competences and with regard for appropriate balance.

This report has been reviewed by a group other than the authors according to procedures approved by a Report Review Committee consisting of members of the National Academy of Sciences, the National Academy of Engineering, and the Institute of Medicine.

This material is based on work supported by the U.S. Department of Agriculture, Agricultural Research Service, under Agreement No. 59-0700-3-078; by the U.S. Nuclear Regulatory Commission under Award No. NRC-04-93-070 (opinions, findings, conclusions, and recommendations expressed herein are those of the authors and do not necessarily reflect the views of the Nuclear Regulatory Commission); and by the U.S. Departments of Defense, Energy, and Health and Human Services; the U.S. Environmental Protection Agency; the American Industrial Health Council; and the Electric Power Research Institute.

Library of Congress Cataloging-in-Publication Data

Understanding risk : information decisions in a democratic society /
    Paul C. Stern and Harvey V. Fineberg, editors.
            p.   cm.
    "Committee on Risk Characterization. Commission on Behavioral and
Social Sciences and Education. National Research Council."
    Includes bibliographical references and index.
    ISBN 0-309-05396-X
    1. Risk assessment. 2. Policy sciences. I. Stern, Paul C.,
1944- . II. Fineberg, Harvey V.   III. National Research Council
(U.S.). Committee on Risk Characterization. IV. National Research
Council (U.S.). Commission on Behavioral and Social Sciences and
Education.
HM256.U53   1996
302'.12—dc20   96-16152
                                                                                    CIP

Understanding Risk: Informing Decisions in a Democratic Society is available for sale from the National Academy Press, 2101 Constitution Avenue, N.W., Box 285, Washington, D.C. 20055. Call 800-624-6242 or 202-334-3313 (in the Washington Metropolitan Area).

COMMITTEE ON RISK CHARACTERIZATION

HARVEY V. FINEBERG (*Chair*), Harvard School of Public Health
JOHN AHEARNE, Sanford Institute of Public Policy, Duke University, and Sigma Xi Center, North Carolina
THOMAS BURKE, School of Hygiene and Public Health, Johns Hopkins University
CARON CHESS, Center for Environmental Communication, Rutgers University
BRENDA DAVIS, Johnson & Johnson Health Care Systems, Inc., Piscataway, New Jersey
PETER DEFUR, Environmental Defense Fund, Washington, D.C.
JEFFREY HARRIS, Department of Economics, Massachusetts Institute of Technology
MARK HARWELL, Rosensteil School of Marine and Atmospheric Science, University of Miami
SHEILA JASANOFF, Department of Science and Technology Studies, Cornell University
JAMES LAMB, Jellinek, Schwartz & Connolly, Washington, D.C.
D. WARNER NORTH, Decision Focus, Inc., Mountain View, California, and Department of Engineering-Economic Systems, Stanford University
KRISTIN SHRADER-FRECHETTE, Department of Philosophy and Program in Environmental Sciences and Policy, University of South Florida
PAUL SLOVIC, Decision Research, Eugene, Oregon, and Department of Psychology, University of Oregon
MITCHELL SMALL, Departments of Civil and Environmental Engineering and Department of Engineering and Public Policy, Carnegie Mellon University
ELAINE VAUGHAN, School of Social Ecology, University of California, Irvine
JAMES WILSON, Resources for the Future, Washington, D.C.
LAUREN ZEISE, California Environmental Protection Agency, Berkeley

PAUL C. STERN, *Study Director*
SARAH CONNICK, *Senior Staff Officer*
THOMAS WEBLER, *Consultant*
MARY E. THOMAS, *Senior Program Associate*

iii

Acknowledgments

Throughout our work together, the committee benefited from the efforts of an extremely dedicated staff. As project director, Paul Stern often took the lead in reconciling and expressing the views of committee members and played a substantial role in crafting this report. Tom Webler, consultant to the committee, contributed his ideas and valuable draft materials. Eugenia Grohman, associate director of reports of the Commission on Behavioral and Social Sciences and Education, lent encouragement, helped streamline our report, and made it more readable. Mary Thomas ably managed our logistic arrangements and communications. Sarah Connick provided valuable assistance in getting the committee started on its work.

The committee was aided in its deliberations by the testimony and advice of many knowledgeable and experienced individuals. The committee acknowledges with appreciation their presentations to the committee:

Alwynelle Ahl, Agricultural and Plant Health Inspection Service,
    U.S. Department of Agriculture
Calvin Bey, Forest Service, U.S. Department of Agriculture
Elinor Blake, Executive Assistant, Hazardous Material Commission,
    Contra Costa County, California
Michael Brody, Office of Policy, Planning and Evaluation, U.S.
    Environmental Protection Agency
Joseph Catruvo, U.S. Environmental Protection Agency

Mark Cunningham, U.S. Nuclear Regulatory Commission

Lynn Desautels, Office of Policy, Planning and Evaluation, U.S. Environmental Protection Agency

Adam Finkel, Resources for the Future

Michael Firestone, Office of Prevention, Pesticides and Toxic Substances, U.S. Environmental Protection Agency

George Fries, Agricultural Research Service, U.S. Department of Agriculture

Philip Harter, attorney, Washington, D.C.

Carol Henry, U.S. Department of Energy

Gordon Hester, Electric Power Research Institute

Karen L. Hulebak, U.S. Food and Drug Administration

Carolyn Leep, Chemical Manufacturers Association

Ray Kent, U.S. Environmental Protection Agency

Carl Mazza, Office of Air and Radiation, U.S. Environmental Protection Agency

Hugh McKinnon, U.S. Environmental Protection Agency

Michael Pompili, Assistant Commissioner of Health, Columbus, Ohio

Greg Schirm, Director, Delaware Valley Toxics Coalition

Donald Stevenson, American Industrial Health Council

We also thank John Lathrop of Strategic Insights for his contributions to the case study on the Florida Power Corporation (in Appendix A).

Erratum

Page ii, line 15 of the Notice, and page x, line 21, of the Preface:

Add the Chemical Manufacturers Association to the list of organizations supporting this study.

Understanding Risk:
Informing Decisions in a Democratic Society
National Academy Press, Washington, D.C.
1996
ISBN 0-309-05396-X

Contents

Preface

For decades the National Research Council has been called on to consider how to improve decisions about risks to public health, safety, and environmental quality. The Research Council has responded with a series of studies that reflect the history of thinking about how society can understand and cope with those risks. *Risk Assessment in the Federal Government: Managing the Process* reported the results of a study that sought "institutional mechanisms that best foster a constructive partnership between science and government" for informing contentious public decisions about hazards to human health from exposures to toxic substances (National Research Council, 1983:1). The study is best known for popularizing the distinction between risk assessment and risk management and raising the issue of how best to keep these functions separate, yet coordinated.

Several years later, *Improving Risk Communication* focused on the relationship between producers and users of scientific information about risks, addressing ways to improve communication "in the service of public understanding and better-informed individual and social choice" (National Research Council, 1989:x). More recently, *Building Consensus Through Risk Assessment and Management of the Department of Energy's Environmental Remediation Program* considered links between risk assessment and public participation. It sought ways to "conduct a credible risk assessment of all the risks at all the sites [where the Department was making restoration after use in the nuclear weapons program], with active participation of all the local participants" (National Research Council,

1994b:vii). At the same time, *Science and Judgment in Risk Assessment* addressed the generic problem of establishing and, when appropriate, changing guidelines for assessing human health risks in ways that deal appropriately with the uncertainties of existing knowledge and the needs of decision makers (National Research Council, 1994a).

Like these previous studies, the present one addresses a broad issue linking risk science and policy. The initial charge formulated the problem as follows:

> The way the nation handles risk often breaks down at the stage of "risk characterization," when the information in a risk assessment is translated into a form usable by a risk manager, individual decision maker, or the public. Oversimplifying the science or skewing the results through selectivity can lead to the inappropriate use of scientific information in risk management decisions, but providing full information, if it does not address key concerns of the intended audience, can undermine that audience's trust in the risk analysis.

This problem was of sufficiently broad interest that the study received support from the U.S. Departments of Defense, Health and Human Services, Agriculture, and Energy, the U.S. Environmental Protection Agency (EPA), the U.S. Nuclear Regulatory Commission, the American Industrial Health Council, and the Electric Power Research Institute. In some of the departments and agencies, the interest and support came from several major internal units. Thus, we were asked to address concerns of entities as diverse as the Centers for Disease Control and Prevention, the Agricultural and Plant Health Inspection Service, civilian and defense organizations responsible for radioactive waste management, the Food and Drug Administration, and EPA's Office of Prevention, Pesticides, and Toxic Substances.

To carry out this broad task, the Research Council convened a committee of 17 members from a variety of specialties including risk assessment, epidemiology, toxicology, ecology, public policy, economics, decision science, social science, medicine, public health, and law. Members were selected to ensure that the perspectives of federal and state regulatory agencies, industry, and environmental and citizens groups would be included, along with those of scientists. And members were selected so as to assure a flexible view of the charge and to provide an overall balance to the committee. Biographical sketches are provided in Appendix C.

At its initial meetings the committee heard from each of its sponsors and considered a detailed letter from representatives of most of the sponsoring agencies that presented a considerably broader reading of the charge, which appears to restrict "risk characterization" to the translation of scientific information already available from risk assessments. In particular, the letter called on the committee to "consider the appropriateness of including in risk characterizations" such considerations as "economic factors, equity issues, risk mitigation and tradeoffs, and technical control feasibility," as well as "environmental-equity issues and other issues of social context," considerations not normally included in risk assessments. The letter also called on the committee for "guidance . . . to improve the dialogue between risk assessors and risk managers prior to and during the development of a comprehensive assessment so that policy and management concerns are understood by all parties." This request implicitly recognized the importance to risk characterization of communication before and during the process of conducting risk assessments, not only after they are complete. Some of the sponsors, particularly the Department of Energy, also indicated that concerns about improving public participation, building trust, and similar issues were among those that had led them to support the study.

As a result of discussion of these concerns with the sponsors' representatives, the committee adopted a revised task statement that reflected a broader charge:

> "Risk characterization" is a complex and often controversial activity that is both a product of analysis and dependent on the processes of defining and conducting analysis. The study committee will assess opportunities to improve the characterization of risk so as to better inform decision making and resolution of controversies over risk. The study will address: technical issues such as the representation of uncertainty; issues relating to translating the outputs of conventional risk analysis into nontechnical language; and social, behavioral, economic, and ethical aspects of risk that are relevant to the content or process of risk characterization.

This charge makes explicit that the committee would consider both translation issues and those processes that determine whether risk characterizations ultimately better inform decision making. The revised charge represents the first step in defining the committee's view of its topic that is reflected in the use of the term "understanding risk" in the title of this volume.

The committee held an informal meeting in March 1994 and six meetings between May 1994 and June 1995 to gather and consider information and to write its report. It engaged in discussions with sponsors' representatives and a variety of outside scientists and risk practitioners whose experiences with risk characterizations the committee believed would be instructive. It sought knowledge from various sources, including experimental research on risk perception and methods of summarizing risk information; studies that evaluate the effects and outcomes of various ways of analyzing and deliberating about risk; and the reflections of experienced practitioners of risk assessment, characterization, and decision making. The committee discussed a wide range of risks, including risks to human health and safety, the environment, and ecosystems and risks from chemicals, foods, ionizing radiation, electromagnetic fields, people's own behavior, exotic organisms or biological materials, and global climatic change. It discussed a wide range of uses for risk characterization, including: informing regulatory decisions on approving drugs, chemicals, and vaccines; setting chemical exposure standards; setting priorities for public expenditures on risk reduction; informing populations at risk from hazardous substances, infectious disease, or their own behavior; and informing legislative debates.

Given the variety of sponsors, risks, and decision situations, the committee emphasized broad considerations about risk characterization rather than those that are specific to certain risks, decision types, or government agencies. It developed consensus about how to think about and organize risk characterization efforts, without trying to offer detailed guidance for particular decision contexts. While reviewing comments on its draft report, the committee learned that the congressionally mandated Commission on Risk Assessment and Risk Management will propose a framework that similarly emphasizes the importance of coupling analysis with the participation of interested and affected parties. The committee welcomes this reinforcement and views its main ideas and conclusions as building on the foundation of previous efforts, including the Research Council reports mentioned above and the work of many others to improve ways of coping with risk situations. If its recommendations can be implemented with appropriate deliberation and judgment, the committee believes that more understandable, scientifically sound, and acceptable decisions will result.

The committee stresses to the readers of this report our conviction that no set of guidelines or procedures can ever substitute for scientific rigor, fairness, and flexibility in coping with dynamic risk situations. Yet we do hope our findings and recommendations will aid those of good will to make sounder decisions about risks.

HARVEY V. FINEBERG, *Chair*
Committee on Risk Characterization

Understanding
RISK

Summary

Coping with risk situations can be complex and controversial. Government and industry have devoted considerable resources to developing and applying techniques of risk analysis and risk characterization in order to make better informed and more trustworthy decisions about hazards to human health, welfare, and the environment, yet these methods often fail to meet expectations that they can improve decision making. One reason lies in inadequacies in the techniques available for analyzing risks. A second is the fundamental and continuing uncertainty in information about risks. Another, less well appreciated, reason for the failure lies in a basic misconception of risk characterization.

Risk characterization is often conceived as a summary or translation of the results of technical analysis for the use of a decision maker. Seen in this light, a risk characterization may fail for two reasons: it may portray the scientific and technical information in a way that leads to an unwise decision, or it may provide scientific and technical information in a way that is not useful for the decision maker. Although such failures do occur, an often overlooked danger to risk decision making is a fundamental misconception about how risk characterization should relate to the overall process of comprehending and dealing with risk.

We propose that it is necessary to reconceive risk characterization in order to increase the likelihood of achieving sound and acceptable decisions. We envision a process in which the characterization of risk emerges from a combination of analysis and deliberation. We offer seven principles for implementing the process.

Risk characterization should be a *decision-driven activity*, directed toward informing choices and solving problems.

The view of risk characterization as a translation or summary is seriously deficient. What is needed for successful characterization of risk must be considered at the very beginning of the process of developing decision-relevant understanding. Risk characterization should not be an activity added at the end of risk analysis; rather, its needs should largely determine the scope and nature of risk analysis.

The aim of risk characterization, and therefore of the analytic-deliberative process on which it is based, is to describe a potentially hazardous situation in as accurate, thorough, and decision-relevant a manner as possible, addressing the significant concerns of the interested and affected parties, and to make this information understandable and accessible to public officials and to the parties.

Although risk characterizations are often completed for the benefit only of an organization's decision maker, it is important to recognize that various other parties use them when they exercise their rights to participate in the decision, either before or after the organization acts. These interested and affected parties include legislators, judges, industry groups, environmentalists, citizens' groups, and a variety of others. Acceptance of risk decisions by the broad spectrum of the interested and affected parties is usually critical to their implementation. Risk characterization processes and products should provide all the decision participants with the information they need to make informed choices, in the form in which they need it. A risk characterization that fails to address their questions is likely to be criticized as irrelevant or incompetent, regardless of how carefully it addresses the questions it selects for attention.

The appropriate level of effort for a risk characterization is situation specific. Judgment is critical in determining the amount, content, and timing of effort that are appropriate for supporting a particular risk characterization. Two things are critical: careful diagnosis of the decision situation to arrive at preliminary judgments and openness to reconsidering those judgments as the process moves along. The procedures that govern risk characterization should leave enough flexibility to be expanded or simplified to suit the needs of the decision.

Coping with a risk situation requires a *broad understanding* of the relevant losses, harms, or consequences to the interested and affected parties.

A risk characterization must address what the interested and affected parties believe to be at risk in the particular situation, and it must incorporate their perspectives and specialized knowledge. It may need to consider alternative sets of assumptions that may lead to divergent estimates of risk; to address social, economic, ecological, and ethical outcomes as well as consequences for human health and safety; and to consider outcomes for particular populations in addition to risks to whole populations, maximally exposed individuals, or other standard affected groups. Under certain conditions, such as when the stakes are high and trust in the responsible organization is low, the organization may need to make special efforts to ensure that the interested and affected parties accept key underlying assumptions about risk-generating processes and risk estimation methods as reasonable.

Adequate risk analysis and characterization thus depend on incorporating the perspectives and knowledge of the interested and affected parties from the earliest phases of the effort to understand the risks. The challenges of asking the right questions, making the appropriate assumptions, and finding the right ways to summarize information can be met by designing processes that pay appropriate attention to each of these judgments, inform them with the best available knowledge and the perspectives of the spectrum of decision participants, and make the choices through a process that those parties trust.

Risk characterization is the outcome of an *analytic-deliberative process*. Its success depends critically on systematic analysis that is appropriate to the problem, responds to the needs of the interested and affected parties, and treats uncertainties of importance to the decision problem in a comprehensible way. Success also depends on deliberations that formulate the decision problem, guide analysis to improve decision participants' understanding, seek the meaning of analytic findings and uncertainties, and improve the ability of interested and affected parties to participate effectively in the risk decision process. The process must have an appropriately diverse participation or representation of the spectrum of interested and affected parties, of decision makers, and of specialists in risk analysis, at each step.

Analysis and deliberation can be thought of as two complementary approaches to gaining knowledge about the world, forming understandings on the basis of knowledge, and reaching agreement among people. *Analysis* uses rigorous, replicable methods, evaluated under the agreed protocols of an expert community—such as those of disciplines in the natural, social, or decision sciences, as well as mathematics, logic, and law—to arrive at answers to factual questions. *Deliberation* is any formal or informal process for communication and collective consideration of issues. Participants in deliberation discuss, ponder, exchange observations and views, reflect upon information and judgments concerning matters of mutual interest, and attempt to persuade each other. Government agencies should start from the presumption that both analysis and deliberation will be needed at each step leading to a risk characterization.

Deliberation is important at each step of the process that informs risk decisions, such as deciding which harms to analyze and how to describe scientific uncertainty and disagreement. Appropriately structured deliberation contributes to sound analysis by adding knowledge and perspectives that improve understanding and contributes to the acceptability of risk characterization by addressing potentially sensitive procedural concerns.

Deliberation needs to be broader and more extensive for some decisions and at some steps than others. It should have, in addition to the involvement of appropriate policy makers and scientific and technical specialists, sufficiently diverse participation from across the spectrum of interested and affected parties to ensure that the important, decision-relevant knowledge enters the process, that the important perspectives are considered, and that the parties' legitimate concerns about the inclusiveness and openness of the process are addressed.

Organizing appropriately broad deliberation presents significant challenges, including managing scarce resources, setting realistic expectations, identifying all the parties that should be involved, and nurturing the process. On the basis of limited research on deliberative methods, we can specify four guidelines.

First, although potentially more time-consuming and cumbersome in the near term, it is often wiser to err on the side of too-broad rather than too-narrow participation. Organizations should seriously assess the need for involvement of the spectrum of interested and affected parties at each step, with a presumption in favor of involvement. If some parties that are unorganized, inexperienced in regulatory policy, or unfamiliar with risk-related science are particularly at risk and may have critical information about the risk situation, it is worthwhile for responsible organizations to arrange for technical assistance to be provided to them from sources that they trust. Broad participation is often needed "early" in the process, and especially in problem formulation.

Second, the conveners of deliberative processes should clearly and explicitly inform participants at the outset about the legal, budgetary, or other external constraints likely to affect the extent of deliberation possible or how the input from deliberation will be used.

Third, deliberative processes should strive for fairness in selecting participants and in providing, as appropriate, access to expertise, information, and other resources for parties that normally lack these resources.

Fourth, managers should build flexibility into deliberative processes, including procedures for responding to requests to reconsider past decisions or to change procedures within externally established limits of time or resources. It must be recognized that even when successful, deliberation cannot be expected to end all controversy. It will not guarantee that decision makers will pay attention to deliberation's outcomes, prevent dissatisfied parties from seeking to delay or override the process, or redress the situation in which legal guidelines mandate that decisions be based on a different set of considerations from those that participants believe appropriate.

Analysis is the best source of reliable, replicable information about hazards and exposures, and as such it is essential for good risk characterization. Relevant analysis, in quantitative or qualitative form, strengthens the knowledge base for deliberations: without good analysis, deliberative processes can arrive at agreements that are unwise or not feasible. The chief challenges are to follow in practice analytic principles that are widely accepted and to recognize the limitations of analysis.

Much attention has been recently given to analytic techniques for benefit-cost analysis and for making quantitative risk comparisons that attempt to reduce many dimensions of risk to one as an aid to decision making. These techniques necessarily simplify real-world situations and require value choices among dimensions of risk. Value judgments that are left implicit in analytic techniques and that are made without broad-based deliberation can cause many difficulties. The key to successful use of these techniques is that a broadly based deliberative process helps shape the analysis, determining which particular techniques are used, and how their results are interpreted.

Much attention has been given to quantitative, analytic procedures for describing uncertainty in risk characterizations. Participants in decisions need to consider both the magnitude of uncertainty and its sources and character: whether it is due to inherent randomness or to lack of knowledge; and whether it is recognized and quantifiable, recognized and indeterminate; or perhaps unrecognized. Unfortunately, the unrecognized sources of uncertainty—surprise and fundamental ignorance about the basic processes that drive risk—are often important sources of uncertainty, and formal analysis may not help if they are too large. Thus, uncertainty analysis should be conducted with care and in conjunction with deliberation and in full awareness of its limitations, especially in the face of unrecognized sources of uncertainty. It is best to focus on uncertainties that matter most to ongoing processes of deliberation and decision. The users of uncertainty analysis should remember that both the analysis and people's interpretations of it can be strongly affected by the social, cultural, and institutional context of the decision.

> The analytic-deliberative process leading to a risk characterization should include early and explicit attention to *problem formulation;* representation of the spectrum of interested and affected parties at this early stage is imperative.

> The analytic-deliberative process should be *mutual and recursive.* Analysis and deliberation are complementary and must be integrated throughout the process leading to risk characterization: deliberation frames analysis, analysis informs deliberation, and the process benefits from feedback between the two.

A recurring criticism of risk characterizations is that the underlying analysis failed to pay adequate attention to questions of central concern to some of the interested and affected parties. This is not so much a failure of analysis as a failure to integrate it with broadly based deliberation: the analysis was not framed by adequate understanding about what should be analyzed. Organizations need to be creative in integrating these two processes. Although a very broad analytic-deliberative process will be appropriate in relatively few instances, those instances have an importance disproportionate to their number. Moreover, it is not always evident in advance whether a risk characterization will require extensive deliberation, integrated with analysis.

A key practical problem for organizations is resolving the tension between the desire for more analysis and deliberation and the need to reach closure. Reaching closure is likely to be most difficult when interests are in strong opposition, when the number of participants is large, and when differences are based on fundamental values. Organizations should consider having the participants in a deliberation adopt procedural rules that enable closure even when substantial disagreements exist. They should also consider two reasons to delay closure: to allow all parties to hear others and be heard and to bring to the surface additional information and concerns that will need to be considered.

Structuring an effective analytic-deliberative process for informing a risk decision is not a matter for a recipe. Every step involves judgment, and the right choices are situation dependent. Still, it is possible to identify objectives that also serve as criteria for judging success:

* *Getting the science right:* The underlying analysis meets high scientific standards in terms of measurement, analytic methods, data bases used, plausibility of assumptions, and respectfulness of both the magni-

tude and the character of uncertainty, taking into consideration limitations that may have been placed on the analysis because of the level of effort judged appropriate for informing the decision.

• *Getting the right science:* The analysis has addressed the significant risk-related concerns of public officials and the spectrum of interested and affected parties, such as risks to health, economic well-being, and ecological and social values, with analytic priorities having been set so as to emphasize the issues most relevant to the decision.

• *Getting the right participation:* The analytic-deliberative process has had sufficiently broad participation to ensure that the important, decision-relevant information enters the process, that all important perspectives are considered, and that the parties' legitimate concerns about inclusiveness and openness are met.

• *Getting the participation right:* The analytic-deliberative process satisfies the decision makers and interested and affected parties that it is responsive to their needs: that their information, viewpoints, and concerns have been adequately represented and taken into account; that they have been adequately consulted; and that their participation has been able to affect the way risk problems are defined and understood.

• *Developing an accurate, balanced, and informative synthesis:* The risk characterization presents the state of knowledge, uncertainty, and disagreement about the risk situation to reflect the range of relevant knowledge and perspectives and satisfies the parties to a decision that they have been adequately informed within the limits of available knowledge. An accurate and balanced synthesis treats the limits of scientific knowledge (i.e., the various kinds of uncertainty, indeterminacy, and ignorance) with an appropriate mixture of analytic and deliberative techniques.

These criteria are related. To be decision-relevant, risk characterization must be accurate, balanced, and informative. This requires getting the science right and getting the right science. Participation helps ask the right questions of the science, check the plausibility of assumptions, and ensure that any synthesis is both balanced and informative.

> **Those responsible for a risk characterization should begin by developing a provisional *diagnosis of the decision situation* so that they can better match the analytic-deliberative process leading to the characterization to the needs of the decision, particularly in terms of level and intensity of effort and representation of parties.**

An agency or organization responsible for risk characterization begins with a diagnosis—explicit or implicit—that includes, at minimum, ideas about the nature of a hazard situation, the purposes for which risk characterization will be used, the kinds of information that will probably be needed, and the kind of decision to be made. Diagnosis should be conducted explicitly far more often than is current practice. Diagnosis begins with surveying what we call the risk-decision landscape, to see what decisions will need to be made. Risk characterization requires different kinds of effort for different categories of decisions. For instance, unique and wide-impact decisions tend to create strong needs for breadth, inclusion, and attention to process; in contrast, for many routine, narrow-impact decisions, a simple, generic risk characterization procedure may suffice. Decisions to simplify should be taken with care, however, because an inappropriate or inflexible decision to use a narrow, routinized, or nonparticipatory process for risk characterization can undermine the decision-making process.

Diagnosis of risk decision situations should follow eight steps: diagnose the kind of risk and the state of knowledge, describe the legal mandate, describe the purpose of the risk decision, describe the affected parties and anticipate public reactions, estimate resource needs and timetable, plan for organizational needs, develop a preliminary process design, and summarize and discuss the diagnosis within the responsible organization. Diagnosis should result in a commitment within the responsible organization about the nature and level of effort of the analytic-deliberative process leading to a risk characterization. Officials of the responsible organization should, however, treat the diagnosis as tentative and remain open to change, always keeping in mind that their goal is a process that leads to a useful and credible risk characterization.

Each organization responsible for making risk decisions should work to *build organizational capability* to conform to the principles of sound risk characterization. At a minimum, it should pay attention to organizational changes and staff training efforts that might be required, to ways of improving practice by learning from experience, and to both costs and benefits in terms of the organization's mission and budget.

These principles may be difficult to follow, particularly with respect to increasing input from some interested and affected parties, involving nonscientists in deliberations about risk analysis, broadening the range of adverse outcomes to consider in risk analysis, and more fully integrating analysis and deliberation, all of which may appear to prolong the decision process or increase its complexity. While we are sensitive to concerns about cost and delay, we note that huge costs and delays have sometimes resulted when a risk situation was inadequately diagnosed, a problem misformulated, key interested and affected parties did not participate, or analysis proceeded unintegrated with deliberation. We believe that following the above principles can reduce delays and costs as much as or more than it increases them.

It is beneficial over time for an organization to use a broad analytic-deliberative process to get the characterization right the first time—to accept immediate costs to avoid greater future costs. We recognize that parties dissatisfied with a risk characterization or risk decision may sometimes seek redress through court challenges or other means. This is to be expected in a democracy, although it adds expense and may constrain efforts to involve the full range of interested and affected parties.

It is critical for the organizations responsible for characterizing risk to have the capability to organize a full range of analytic-deliberative processes, including the broadly participatory ones that some risk situations warrant. It is also critical that they develop the capability to cope with attempts by some interested and affected parties to delay decision, and to develop a range of strategies for reaching closure. To these ends, each organization responsible for risk characterization should consider making special efforts in training staff; acquiring analytic expertise with regard to ecological, social, economic, or ethical outcomes; and making organizational changes to improve communication across subunits and to allow for the flexibility and judgment necessary to match the process to the decision.

Every organization should implement explicit practices to promote systematic learning from its efforts to inform and make risk decisions, so as to improve analytic-deliberative processes. It should work with the interested and affected parties to define criteria for evaluating these processes. It should devise systems of evaluation and feedback to allow for mid-course corrections that save time and money, pretesting of materials summarizing risk information, and the use of retrospective analysis to improve future efforts. In addition, institutions that provide scientific support for these organizations, such as federal scientific agencies and industry-based research institutes, should support systematic efforts that build knowledge about analytic-deliberative processes and that may have general value for many organizations.

Evaluation or feedback should take a form appropriate to the scale and nature of the analytic-deliberative process. Evaluation is important both during and after the process. It can use a variety of formal and informal methods, including surveys, experimental tests of informational materials, evaluation research methods, simulations, quasi-experimental evaluations of new procedures, feedback from broadly based advisory groups that review past practice, and systematic case study research on libraries of case files.

An expanded concept of risk characterization raises legitimate questions about practicality, such as whether it would unacceptably increase the costs and time for making decisions and whether any increased costs would lead to better or more acceptable decisions. These are reasonable concerns, but we believe, on balance, that the process we propose is likely to improve outcomes. Experience shows that analyses, no matter how thorough, that do not address the decision-relevant questions, use reasonable assumptions, and meaningfully include the key affected parties can result in huge expenses and long delays and jeopardize the quality of understanding and the acceptability of the final decisions. These dangers associated with past approaches to risk characterization are sufficient in our judgment to warrant making a serious trial of the broader concept. We also emphasize that the approach we propose expands the process only as appropriate to specific situations: For many risk issues, relatively little change will be needed in risk characterization; for some of society's most important risk issues, however, a broad and extensive analytic-deliberative process can lead to better informed and more widely acceptable decisions.

1

The Idea of Risk Characterization

$$D$$uring the past several decades
many areas of government policy associated with hazards to health,
safety, and the environment have become increasingly contentious. Despite much new legislation and extensive efforts by the agencies charged
with implementing the legislation, dissatisfaction and controversy continue. A continuing debate on regulatory reform has not yet reached
consensus on how governmental institutions and procedures should be
structured to make decicions better and more broadly acceptable. Many
believe that increased use of risk analysis[1] is appropriate. The expectation that clear and concise characterizations of existing information about
risks, costs, and benefits will lead to informed and acceptable regulatory
decisions is attractive; it may, however, be naive. One reason lies in
inadequacies of the techniques available for risk analysis. A second is the
fundamental and continuing uncertainty in information about risks. Another, less well appreciated reason lies in a basic misconception of risk
characterization and its relation to the overall process of comprehending
and dealing with risk. Risk characterization involves complex, value-
laden judgments and a need for effective dialogue between technical experts and interested and affected citizens who may lack technical expertise, yet have essential information and often hold strong views and
substantial power in our democratic society.

---

[1]See Glossary for the terms used in this volume.

We believe the iterative analytical-deliberative process described in this volume holds much promise for improving risk characterization, informing decisions, and making those decisions more acceptable to interested and affected parties. The technical and analytical aspects of risk analysis must be balanced with a concern for appropriate involvement by interested and affected parties in all steps of the decision-making process, including those leading to risk characterization. Analysis and citizen involvement are not separate steps to be carried out in sequence, but must be combined into an effective synthesis. Our approach involves a substantial change from the formulation of risk analysis that many federal agencies and other organizations have been using for more than a decade.

Many groups before us have studied aspects of risk decision-making processes in order to improve decisions. We undertook this study to evaluate and make recommendations about risk characterization, described in the committee's initial task statement as the part of the decision process at which "the information in a risk assessment is translated into a form usable by a risk manager, individual decision maker, or the public." Stating the committee's charge in this way highlights a central dilemma of risk decision making in a democracy: detailed scientific and technical information is essential for understanding risks and making wise decisions about them, yet the people responsible for making the decisions and the people affected by the decisions and who may therefore also take part in them are not themselves expert in the relevant science and technology.

This dilemma has spawned numerous technical and policy attempts to resolve it. It is important to recognize, however, that in many areas of hazard management, the dilemma has not posed much of a problem. For example, airline and automotive safety have improved fairly continuously over long periods, and the public has trusted the responsible institutions to monitor and maintain safety, even though few citizens understand the technologies. The same has been generally true for the safety of foods and drugs, earthquake engineering, and the issuance of some routine environmental permits. In other areas, however, the dilemma has been stark. In the management of radioactive waste, toxic chemicals, and hazardous industrial facilities, for instance, technical experts and the responsible agencies are mistrusted by many of those who participate in risk decisions. Our concern is especially focused on decisions in which, as in the latter set of cases, participants are likely to come into conflict about the adequacy of scientific knowledge; about issues of fairness, access, and consent in the decision process; or about basic goals and values. Such decisions are relatively few in number, but usually great in importance. Moreover, decisions that have not aroused this sort of contention in the past could do so in the future.

We have focused narrowly on one aspect of risk decision making—the problem of characterizing risk so that better informed decisions can be made—but in other respects, we have taken a very broad view. First, we have not restricted ourselves to a particular kind of hazard or risk. Thus, the issues we discuss are relevant whether the goal is to understand the risks of cancer to humans, noncancer health risks, or risks to ecological, social, or political systems. We believe the principles we develop are applicable over a very broad range of hazards and risks.[2]

Second, we do not restrict our discussion to particular kinds of organizations that may engage in characterizing risks. For specificity, we sometimes write as if the responsible organization is a federal, state, or local government agency, and in some places, the language may seem to be addressed even more specifically to federal regulatory agencies. We certainly intend that what we say will meet the needs of such organizations. We believe that it also has wider applicability to public health organizations, industrial organizations, nonprofit organizations, and others who prepare descriptions of risks to health, safety, or the environment. It may be, however, that adversarial settings such as courts do not place the same demands on participants because those settings provide their own mechanisms for including different perspectives.

Third, we find that in order to improve risk characterization, one must consider other parts of the risk decision process, particularly the various analytic activities that provide the information used in characterizing risks. The purpose of risk characterization is to improve understanding of risk, and everything that goes into such understanding is necessary for effective risk characterization. We develop this theme in detail in this chapter.

Finally, we offer a strategic approach and a set of principles for a better understanding of risk, rather than a set of guidelines or procedures that can be applied in a routinized way for all risk situations. Our recommended strategy implies the need for very extensive effort and expense for characterizing some risks, but relatively little in most instances: the level and kind of effort required are highly situation-specific. Thus, it is imperative for organizations engaged in characterizing risk to consider the situation carefully at the outset and to build flexibility into their char-

---

[2]Similarly, the issues we discuss are relevant regardless of whether decisions are normally based on formal risk estimates (as is typically the case with cancer risks to humans); surrogate measuress, such as a dose observed to have no significant effect in laboratory animal studies, modified by a margin of safety (as is commonly done with noncancer health risks to humans); or made without reference to standard techniques of risk estimation or standard decision-making rules (as with many risks to ecological, social, and political systems).

acterization procedures to allow for unexpected situations. It is no doubt possible to devise more detailed guidelines, consistent with our principles, for risk characterization in particular kinds of situations. We believe that the most useful contribution we can make in this volume is to propose a broadly applicable strategy and conception that allows organizations to deal with some of the most prevalent and serious challenges of risk characterization.

BEYOND TRANSLATION

The committee's initial charge presumed that the key to resolving the dilemma is to "translate" scientific knowledge into a form usable by decision makers. We agree that it is essential to make science useful to decision makers, but we have concluded that doing this involves much more than translation: it requires a different view of the purpose and form of risk characterization.

Risk characterization has typically been seen as a summarization of scientific information. This understanding is succinctly stated in *Risk Assessment in the Federal Government: Managing the Process* (National Research Council, 1983:20), widely referred to as the Red Book:

> Risk characterization is the process of estimating the incidence of a health effect under the various conditions of human exposure described in exposure assessment. It is performed by combining the exposure and dose-response assessments. The summary effects of the uncertainties in the preceding steps are described in this step.

Risk characterization is seen here as the final step in the process of risk assessment. It combines the results of a completed hazard identification, exposure assessment, and dose-response assessment into a concise estimate of adverse effect in a given population. Figure 1-1, taken from the Red Book, schematically represents the traditional view of risk characterization and its relations to the other elements of risk decision making. In this schema (National Research Council, 1983:28), risk characterization involves "no additional scientific knowledge or concepts" and only a minimal amount of judgment. This view remains prevalent in federal agencies. For example, a 1992 memorandum by a deputy administrator of the U.S. Environmental Protection Agency (EPA) clearly defines risk characterization as a summarization or translation process coming at the completion of a scientific analysis (Habicht, 1992). It was reaffirmed by the EPA in a 1995 policy statement using almost identical language (Browner, 1995). It has also been applied, with some modification, to the characterization of ecological risks (U.S. Environmental Protection Agency, 1992a). And it is by no means confined to a single agency of the

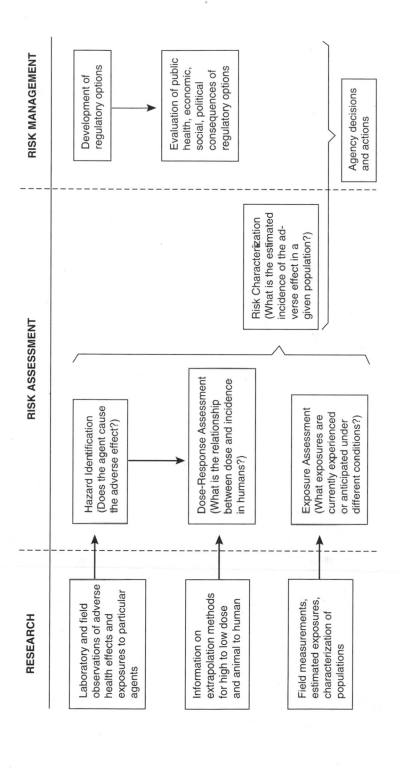

FIGURE 1-1. Elements of risk assessment and risk management, as depicted in *Risk Assessment in the Federal Government* (National Research Council, 1983:28).

government (see, e.g., the special June 1994 issue of *Risk Analysis* on "The Risk Assessment Paradigm After Ten Years").

We have concluded that the view of risk characterization as a summary is seriously deficient, and we propose a more robust construction. Risk characterization must be seen as an integral part of the entire process of risk decision making: what is needed for successful characterization of risk must be considered at the very beginning of the process and must to a great extent drive risk analysis. If a risk characterization is to fulfill its purpose, it must (1) be decision driven, (2) recognize all significant concerns, (3) reflect both analysis and deliberation, with appropriate input from the interested and affected parties, and (4) be appropriate to the decision.[3] The rest of this section describes and illustrates these four facets of risk characterization and the risk decision process.

A Decision-Driven Activity

The purpose of risk characterization is to enhance practical understanding and to illuminate practical choices. A carefully prepared summary of scientific information will not give the participants in a risk decision the understanding they need if that information is not relevant to the decision to be made. It is not sufficient to get the science right; an informed decision also requires getting the right science, that is, directing the scientific effort to the issues most pertinent to the decision.

In 1994 the EPA completed a $6-million scientific reassessment of the health risks of dioxin (U.S. Environmental Protection Agency, 1994a), undertaken to resolve a major controversy about the dose-response relationship between dioxin exposure and possible human health effects, particularly cancer. Under the working assumptions of a linear dose-response relationship for cancer, existing data imply that one form of dioxin, TCDD, is one of the most highly carcinogenic of all chemicals. But analyzing these same data under an alternative hypothesis, dioxin could be much less dangerous. EPA's 96-page risk characterization attempts to synthesize an estimated $1 billion worth of scientific research and a 2,000-page reassessment document. Yet the characterization has not resolved the scientific issues (e.g., Clapp et al., 1995; Environ Dioxin Risk Charac-

---

[3]This idea of risk characterization is the crystallization of much thinking about risk over at least two decades. Much of it is implicit in the common belief among practitioners of risk analysis that it is imperative to engage in repeated interaction with the client; we simply add that risk characterizations have many clients. Intellectual antecedents for many of our ideas can be found in earlier work (e.g., Morse and Kimball, 1951; Howard, 1966, 1968; Morgan, 1981; National Research Council, 1983, 1989, 1994a; Edwards and von Winterfeldt, 1987; Keeney, von Winterfeldt, and Eppel, 1990; von Winterfeldt, 1992).

terization Expert Panel, 1995), and its authors and critics agree that a substantial additional research effort will be necessary if this is to be done.

To what extent would an improved dioxin risk characterization be useful for making decisions, even if scientists could agree on it? Would it be the right science? Among the most likely present uses for such a risk characterization are to inform decisions concerning operating permits for municipal and industrial facilities, siting waste incinerators, making policy decisions about Vietnam veterans exposed to dioxin, and remediating Superfund sites. Yet in all of these contexts, dioxin is only one of many hazardous chemicals involved and cancer is only one of many outcomes of concern, so dioxin-induced cancer is at best only part of the problem and a dioxin risk characterization, though relevant, can only hope to provide some of the information needed for the decision. Moreover, one effect of exclusive or intense attention to quantifying the dioxin-cancer link and its uncertainty may be to draw attention away from other risk-related concerns, which may be more important to participants in the decisions and may require different kinds of analysis. These concerns might include questions about the fairness of exposing a community that may have an abundance of toxic chemical sites to yet another site, about whether the local population has characteristics that make it unusually susceptible to damage from an additional body burden of dioxin, or about effects of the contemplated action on local property values. Some of these (or other) issues may be the most important ones for a particular decision. A risk characterization focused solely on scientific questions about the dose-response relationship of dioxin to cancer may be highly unsatisfactory to some people because it is only marginally relevant to their most serious concerns.

Since the 1950s, national policy on the disposal of high-level radioactive waste from the civilian nuclear power industry has been to store the wastes permanently in deep, underground repositories. Billions of dollars have been spent on studies to characterize the risks of leakage of radioactive materials from proposed repository sites into the environment. Much recent effort has been devoted to what is currently the only proposed high-level repository site, at Yucca Mountain, Nevada. The analyses have repeatedly been criticized on technical grounds, and new technical objections have led to new studies. Although the studies have convinced many members of the technical community that the site is potentially acceptable, most Nevada citizens and many others remain unconvinced. Some of the objections probably result from disagreement about what decision needs to be made.

A major gap in the argument to use the Yucca Mountain site is the lack of a convincing case that a permanent repository is needed now for environmental, safety, or health reasons. Surface storage has been judged by the Nuclear Regulatory Commission to be acceptably safe well into the next century, and "no comprehensive appraisal is now available of the probable costs and risks of continuing the present temporary waste disposal practices" for decades more in comparison with the risks and costs of putting the waste in a permanent repository (National Research Council, 1995:13). In the Yucca Mountain case, the government's risk characterizations seem to have relied too much on only one subset of scientific information, presuming that it was obvious which question needed to be answered. Opponents were concerned with a different set of issues, which were not addressed in the risk analyses, such as the fairness of placing nuclear waste in a region that does not have any nuclear power plants and is already host to the nation's nuclear testing facility (several such concerns are listed in National Research Council, 1995:21-23). Consequently, to many people, characterizations of the Yucca Mountain site are at best irrelevant, and at worst dangerously misleading because they focus attention on the wrong question.

A contrast to these two examples comes from the Man and Biosphere (MAB) Program organized by the U.S. Department of State as part of a larger international effort. In an MAB activity over several years, more than 100 natural and social scientists from various federal and state agencies and from universities have considered policy options for managing surface water so as to maintain a sustainable ecosystem in and around Florida's Everglades (Harwell et al., in press). Changes in the ecosystem and possible responses to them entail risks to endangered species, to drinking water quality in nearby metropolitan areas, and to the livelihoods of sugar growers. The scientists considered all these risks carefully, but from a perspective different from that typical in risk assessments.

They defined the problem not as one of estimating and reducing risks, but as one of developing a shared vision of desired conditions of the ecosystem. They then identified development strategies consistent with such a vision and proposed governance structures that could adaptively manage the social-ecological system as it changed and as new knowledge developed. They considered several scenarios for change in human management of the ecosystem and analyzed them in terms of their compatibility with goals of sustainable economic and social development and with a widely shared vision of ecosystem use. The MAB effort is noteworthy for its problem-driven approach, particularly its extensive and explicit efforts to understand the decisions to be made, rather than presuming that decision makers would gain the understanding they needed from estimates of the ecological, health, and economic costs and benefits of previously defined choices. In fact, the process generated policy options that had not previously been considered and that might be more acceptable, both socially and ecologically, than any that might otherwise have been considered. (A more detailed description of the MAB activity is in Appendix A.)

Recognizing All Significant Concerns

The people who participate in risk decisions—public officials, experts in risk analysis, and interested and affected parties—may be concerned with a variety of possible harms or losses. Sometimes, risks to social, ethical, or ecological values are at least as important as risks to health and safety. The analysis that will be the basis for a risk characterization must pay explicit attention to the breadth of the significant issues. This is often best done by involving the spectrum of decision participants explicitly in formulating the problem to be analyzed.

In recent years a number of states have organized "comparative risk" projects to develop strategies for setting priorities for environmental protection efforts, based on a ranking of risks. Building on the experiences of other states, California began its project in 1992 by providing for input from a wide variety of citizens, which led to considerable elaboration of the issues the project addressed. For example, at the request of participants in the process, an environmental justice committee was created, and it raised some fundamental questions about risk ranking as a strategy. Arguing that risk-based ranking gives insufficient emphasis to community participation, pollution prevention, and the disproportionate risk burdens borne by some communities, the committee proposed giving these three concerns a more important place. The comparative risk project responded by paying increased attention to the distribution of risks to human health and welfare in its analysis. This process—to the satisfaction of some and the consternation of others—initiated broader statewide debate about the goals of environmental policy (California Environmental Protection Agency, 1994; Stone, 1994; also see Appendix A).

Another example of the need to recognize all significant concerns comes from East Liverpool, Ohio, where a hazardous waste incineration project has been controversial for over a decade. Risk assessment studies conducted before, during, and after the construction of the facility have not convinced a number of local constituencies, including public health officials, to accept the incinerator. A series of recent test burns, designed to reduce uncertainty about the risk, only increased the controversy. An underlying cause of the opposition was that people were concerned with issues not addressed in the risk assessments. Some of these, including risks during start-up and shut-down and waste transportation, were eventually incorporated in risk analyses, but several other concerns never were. One of these involved the broader policy issue that approving a waste incinerator might encourage increased production of hazardous waste. Another was that delaying definitive risk analysis until test burns were performed might permit so much investment in construction and testing that government could not then refuse the incinerator, regardless of the results of the risk assessments. Because of this concern, some of the neighbors distrusted the entire risk assessment process. A third concern was with the physical effects of pollutants on the local population, which was exposed to various other industrial emissions. A fourth was with the adequacy of risk assessment methodology for making public health decisions: the local public health department was not satisfied with the "theoretical assumptions" in the risk assessments and wanted decisions to be based on ongoing monitoring of air, soil, and crops. (The East Liverpool incinerator case is described in more detail in Appendix A.)

An Analytic-Deliberative Process

Improving risk characterization requires attention to two discrete but linked processes: analysis and deliberation. Analysis uses rigorous, replicable methods developed by experts to arrive at answers to factual questions. Deliberation uses processes such as discussion, reflection, and persuasion to communicate, raise and collectively consider issues, increase understanding, and arrive at substantive decisions. Deliberation frames analysis and analysis informs deliberation. Thus, risk characterization is the output of a recursive process, not a linear one. Analysis brings new information into the process; deliberation brings new insights, questions, and problem formulations; and the two build on each other. The analytic-deliberative process needs input from the spectrum of interested and affected parties. Four recent cases provide examples of this process.

In 1992 and 1993 the Environmental Protection Agency sponsored a negotiation to set maximum concentration levels for the by-products of chlorine and other disinfectants used to eliminate microbial contamination of drinking water. The agency had signed a consent order setting a 1994 deadline for proposing a rule, but it subsequently concluded that any rule it could propose by that date would be vulnerable to legal challenge because of lack of data. In an attempt to avoid long delays in litigation, EPA invited the interested and affected parties to join it in negotiating a rule. The negotiating group relied on a technical advisory committee to analyze the risks and on processes of dialogue and persuasion to try to reach agreement on a rule.

The negotiators soon ran up against a key information gap: they had little information about the quality of untreated water in different regions, so they could not adequately characterize the risks from microbial pathogens in water left untreated or the level of risk likely to remain after disinfection. Without reliable estimates of these and other factors, they could not be confident about making a rule. The composition of un-treated water was important to some of the interested and affected parties because they wanted to consider rules allowing non-chlorine disinfectant technologies. To consider the feasibility and wisdom of using such tech-nologies would require additional analysis, using new information. The negotiators dealt with the approaching regulatory deadline by proposing two rules: a provisional rule limiting disinfectant by-product concentra-tions and a rule requiring large public water supply systems to collect information on pathogens in source water, disinfectant by-products and their chemical precursors, and other matters. They agreed on a plan for reconsidering and possibly revising the limits in a few years on the basis of the new information collected, which they believed might well justify a revision of the rule. (A description of the regulatory negotiation for disin-fectant by-product concentrations appears in Appendix A.)

Decades of experience with new drugs that came close to approval, only to be found to have unacceptable side effects, had made the Food and Drug Administration (FDA) cautious about approving new clinical trials of compounds that had not received extensive safety and efficacy testing in animals. The epidemic of AIDS (acquired immune deficiency syndrome) and the virus that causes it created strong demand for a faster approval process for new drugs. Although the FDA resisted at first, involvement of AIDS activists and that part of the medical community involved in clinical trials resulted in new, expedited protocols for certain situations. Reexamination of the risk situation led to recognition that some assumptions made for other drug approval decisions did not neces-sarily apply to drugs for AIDS.

In 1989 the Florida Power Corporation began to search for a site for a new 2000-megawatt coal/gas-fired power generation station in its service area. Drawing on the knowledge and judgments of the corporation's technical and managerial staff and the knowledge and judgments of an external environmental advisory group, a consulting team constructed weighted lists of criteria for excluding potential sites. Most of the analysis was done by a consulting firm; most of the deliberation was done by the corporation's staff and the advisory group. Results of each of five rounds of deliberation instructed the consultants on which factors and weights to use in analyzing the potential sites. This iteration between analysis and deliberation was repeated, with the selection criteria becoming more de-fined each time, and more possible sites eliminated from the list. During the fifth and final phase, the six remaining sites were ranked, and one preferred and two alternate sites were selected. As of November 1995 the preferred site was in the licensing stage. (More detail on this example is in Appendix A.)

A rabies outbreak among raccoons in New Jersey and the risk of its spread to humans prompted the creation of an Interagency Rabies Task Force that included not only representatives from several departments of state government and county government, but also a range of outside experts. The task force considered conducting a field trial of air-dropped oral rabies vaccine in the one unaffected area of the state to prevent the spread of disease to raccoons there, but it did not want to proceed without assurances of public acceptance because the test would expose the public to baits containing the vaccine and because property owners would have to be asked to include their properties in the test area. With a survey and public meetings, the task force elicited public concerns and found widespread support for the program. During the field trial, mailings and press releases kept the public informed of progress. Content analysis of newspaper articles and letters and studies of telephone inquiries and complaints also were used to gather data about public response to the drop. After the trial was completed, the Task Force conducted mail and telephone surveys and found broad public support not only for the rabies experiment, but also for the way in which the potential public concerns were handled (Pflugh, no date).

Matching the Process to the Decision

Different kinds of risk decisions require different kinds and levels of analysis and deliberation in support of risk characterization. For a series of similar risk situations, one might establish routines for risk analysis, characterization, and decision making that embody clear and consistent expectations about how the problem is defined, which options are to be considered, what kinds of evidence are to be considered, who is to participate in the process, and so forth. For novel, complex, or highly controversial risk situations—which often involve questions about major potential impacts and the equity of the distribution of risks and benefits—routines are likely not to be satisfactory. It is likely to be necessary to develop unique procedures for characterizing risk in these situations.

Some examples of procedures involving repetitive risk decisions are those for reapproving existing permits for discharge of pollutants from industrial plants, for testing new drugs prior to approval decisions, for issuing premanufacturing approval for the industrial production of new chemicals, and for deciding whether to exclude an individual from receiving a vaccine or giving blood. We do not mean to imply that all the current procedures for these and similar decisions are appropriate; only that it is often appropriate to develop standard procedures. In fact, situations and knowledge change even for routine decisions, and standard procedures for risk analysis and characterization should be reevaluated from time to time.

Risk situations need to be accurately diagnosed to determine whether existing standard procedures should be applied, whether new procedures need to be devised, whether additional information is needed to decide which approach to follow, and what extent and type of analytic and deliberative effort may be needed to come to such decisions. In medicine, experienced clinicians use a combination of knowledge, experience, and judgment to make diagnoses. The situation is closely comparable for those who must diagnose a risk situation and prescribe the appropriate kinds and level of analysis and deliberation needed and the appropriate breadth of participation.

PARTICIPATION AND KNOWLEDGE IN RISK DECISIONS

In the framework we have outlined, risk characterization cannot succeed as an activity added at the end of a risk analysis, but must result from a recursive process that includes problem formulation, analysis, and deliberation. Two essential aspects of that process are appropriately broad participation by the interested and affected parties and appropriate incorporation of science.

Rationales for Participation

There are three compelling rationales for broad participation in risk decisions. They have been classified as normative, substantive, and instrumental (Fiorino, 1990). The normative rationale derives from the principle that government should obtain the consent of the governed. Related to this principle is the idea that citizens have rights to participate meaningfully in public decision making and to be informed about the bases for government decisions. These ideas are embodied in laws, such as the Administrative Procedure Act and the Freedom of Information Act, although these laws and their associated procedures have not always been implemented in ways that involved meaningful participation (e.g., Houghton, 1988; Kathlene and Martin, 1991; Lynn and Busenberg, 1995).

The substantive rationale is that relevant wisdom is not limited to scientific specialists and public officials and that participation by diverse groups and individuals will provide essential information and insights about a risk situation. As we show in detail in Chapter 2, nonspecialists may contribute substantively to risk characterization—for example, by identifying aspects of hazards needing analysis, by raising important questions of fact that scientists have not addressed, and by offering knowledge about specific conditions that can contribute more realistic assumptions for risk analyses. Nonspecialists may also help design decision processes that allow for explicit examination, consideration, and weighing of social, ethical, and political values that cannot be addressed solely by analytic techniques, but also require broadly participatory deliberation.

The instrumental rationale for broad public participation is that it may decrease conflict and increase acceptance of or trust in decisions by government agencies. Mistrust is often at the root of the conflicts that arise over risk analysis in the United States (see, e.g., Bella, 1987; English, 1992; Flynn and Slovic, 1993; Kasperson, Golding, and Tuler, 1992; Laird, 1989; Pijawka and Mushkatel, 1992; Renn and Levine, 1991; Slovic, Flynn, and Layman, 1991; U.S. Department of Energy, 1992). A combination of psychological tendencies to notice, believe, and give more weight to trust-destroying than to trust-building information, and social factors, such as the tendency of mass media to favor bad news and of some special interest groups to encourage distrust to influence policy debates, make trust very fragile (Slovic, 1993a). Some observers have suggested that improving risk analysis and characterization may have little practical effect on public policy without efforts to rebuild trust by improving participation (Kunreuther, Fitzgerald, and Aarts, 1993; Leroy and Nadler, 1993; Slovic, 1993a). Simply providing people an opportunity to learn about the problem, the decision-making process, and the expected benefits of a decision may improve the likelihood that they will support the decision (Peelle, 1979; Peelle et al., 1983; Peelle and Ellis, 1987). Even if participation does not increase support for a decision, it may clear up misunderstandings about the nature of a controversy and the views of various participants. And it may contribute generally to building trust in the process, with benefits for dealing with similar issues in the future.

Role of Science

Reliable technical and scientific input is essential to making sound decisions about risk. Scientific and technical experts bring indispensable substantive knowledge, methodological skills, experience, and judgment to the task of understanding risk.

A few less obvious points are worth emphasizing about the role of scientific analysis in risk decisions. First, such analysis requires contributions from many, diverse disciplines. In particular, risk analysis often needs the substantive and methodological expertise of the economic, social, and behavioral sciences: for instance, effects on property values, tourism, scenic value, human population migrations, fairness, and public trust in government may be important outcomes of risk decisions, and they are in some cases amenable to rigorous scientific analysis. Even health risks cannot be estimated accurately without a good understanding of the behavior of the individuals and organizations that control or are affected by hazardous substances or processes.

Second, in addition to their specialized disciplinary knowledge, scientists bring a capacity to build systematic and reliable ways of analyzing and interpreting information about new situations. As already noted and as elaborated further in Chapter 2, nonspecialists sometimes have information and knowledge to contribute to the risk decision process. It is important to incorporate such knowledge in a valid scientific framework.

Third, scientific analysis may not always be neutral and objective as a decision-making tool, even when it meets all the tests of scientific peer review. Good scientific analysis is neutral in the sense that it does not seek to support or refute the claims of any party in a dispute, and it is objective in the sense that any scientist who knows the rules of observation of the particular field of study can in principle obtain the same results. But science is not necessarily neutral and objective in its ways of framing problems. For example, analyses of the risks of drunk driving that highlight drivers' behavior as a cause of traffic fatalities draw attention away from the equally significant factors of automobile and highway design; analyses of the cancer risks of industrial chemicals divert attention from the possibly comparable risks from naturally occurring chemicals in foods (National Research Council, 1996); and analyses of the risks of indoor air pollution draw attention away from the problems of ambient air pollution—and vice versa. Similarly, analyses of the costs of environmental regulation often serve the policy arguments of the opponents of regulation, while analyses of the risks of unregulated activities bolster the arguments of the proponents of regulation. Each kind of analysis is appropriate by itself, but if the overall scientific effort is tilted too far toward only one of the legitimate formulations of a problem, it tends to yield biased understanding.

Science is not necessarily neutral either, in its choices of assumptions. Analysis is compromised for decision-making purposes when it is based on assumptions about the conditions of hazard exposure that are known to be unreasonable by decision participants who were not consulted when the assumptions were selected. Even standard statistical assumptions can raise questions of bias. The assumption of the null hypothesis as used in risk analysis contains an implicit bias because it places a greater burden of proof on those who would restrict than those who would pursue a hazardous activity, presuming these activities are safe until proven otherwise. Evidence that science has been censored or distorted to favor particular interested parties has long been a source of conflict over risk characterizations (e.g., Rosner and Markowitz, 1985; Lilienfeld, 1991).

Chapter 2 discusses many ways in which judgments made in the course of risk analysis can undermine the quality of risk characterization, even when the analysis meets stringent scientific tests.

Fourth, scientists may be in a specially powerful position to influence decisions because many hazardous substances or activities have non-obvious and delayed effects that can be uncovered and quantified only with highly technical methods. Without specialized skills, nonscientists may be at a disadvantage in trying to confirm or challenge scientists' claims or judgments.

Fifth, science alone can never be an adequate basis for a risk decision. This point deserves special emphasis in the light of recent proposals for "risk-based" decision rules that could tie public risk decisions to standardized technical procedures of risk analysis. Risk decisions are, ultimately, public policy choices. In principle, analysis of a set of alternative decisions could show which would produce the fewest deaths, the fewest new cancers, the fewest workdays lost to illness, or the least cost to a manufacturer under given circumstances, but it cannot tell how these different effects should be weighed in the context of the decision. No amount of analysis can determine whether cancer-incidence rates should be more important to society than the number of workdays lost, or whether preventing cancer should be more important than preventing reproductive disorders, or whether reducing the prevalence of environmental illness in a broad population should be more important than ensuring an equitable distribution of the risk across subpopulations or a reduction of risk to particular subpopulations (e.g., children, the elderly). No amount of analysis can tell whether a 30 percent lifetime cancer risk to one individual is better or worse than a 15 percent risk to two individuals. No amount of analysis can tell whether the loss of one more wetland is more important than the loss of ten jobs, or ten thousand jobs. Analysis can gather useful information about which tradeoffs citizens as individuals would prefer, but scientists cannot and should not be expected to make decisions that involve societal values. A specialist's role is to bring as much relevant knowledge as possible to participants in a decision, whose job is to make the value-laden choices.

These characteristics of science and scientific analysis all show the importance of appropriately broad-based deliberation as well as analysis: to determine what kind of analysis a decision requires; to incorporate information from disparate sources; to determine when analysis is appropriately balanced; and to determine how to synthesize the results of analysis to make them useful to decision participants. Good science is a necessary—in fact, an indispensable—but not sufficient basis for good risk characterization.

AN EXPANDED FRAMEWORK

The aim of risk characterization, and therefore also of the analytic-deliberative process on which it is based, is to describe a potentially hazardous situation in as accurate, thorough, and decision-relevant a manner as possible, addressing the significant concerns of the interested and affected parties, and to make this information understandable and accessible to the parties and to public officials. If the underlying process is unsatisfactory to some or all of the interested and affected parties, the risk characterization will be unsatisfactory as well. A risk characterization can be only as good as the analytic-deliberative process that produces it.

Figure 1-2 presents our conception of the risk decision process. The rest of this chapter specifies the role of risk characterization in this process; the rest of the book elaborates on the key elements in the figure and on ways to improve risk characterization.

A New Definition and Its Implications

We begin with a definition of risk characterization and then elaborate key parts of it:

Risk characterization is a synthesis and summary of information about a potentially hazardous situation that addresses the needs and interests of decision makers and of interested and affected parties. Risk characterization is a prelude to decision making and depends on an iterative, analytic-deliberative process.

A risk characterization has many users. Risk characterization documents are often prepared as if the only significant users will be legally designated decision makers, such as government officials. For example, the 1995 EPA policy statement on risk characterization (Browner, 1995) distinguishes risk characterization from "risk communication" on the grounds of who uses it. Risk communication, according to the statement, "emphasizes the process of exchanging information and opinion with the public," while risk characterization "addresses the interface of risk assessment and risk management." In this usage, a summary of knowledge about risk is a risk characterization if it is addressed to agency officials (it is presumed to be "available to the public" as well), but not if it is intended primarily for others. We do not make this distinction because summaries prepared primarily for internal use may ultimately be used by a range of parties and because such summaries can often benefit from exchanges with "the public" as well as within the agency. Agencies should recognize from the start that a risk characterization should be

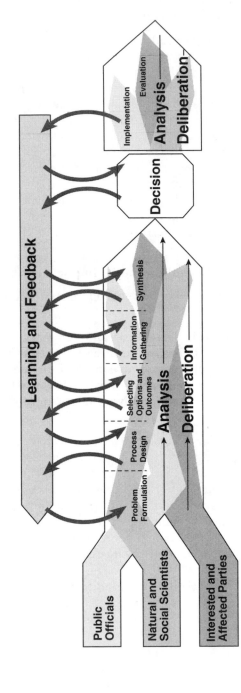

FIGURE 1-2. A schematic representation of the risk decision process.

useful to multiple parties with different interests, concerns, and information needs.

A risk characterization may need to consider a wide variety of outcomes or consequences. In principle, the full range of potential harms and losses from a hazard is appropriate for treatment in a risk characterization. In addition to the biological and physical outcomes that are typically covered, decision makers and interested and affected parties often need to know about the significant economic costs and benefits of alternatives, the secondary effects of hazard events, or the efficacy of alternative regulatory mechanisms. For some decisions, consequences to human health and environmental quality are only part of what is of concern; concerns may also include such matters as geographical, racial, or economic equity, intergenerational tradeoffs, and informed consent by those who will be affected by a decision. Many of these issues can be analyzed systematically, and the summary of such analyses should be included in a risk characterization.

Problem formulation is a paramount consideration. Because a risk characterization is geared to a risk decision, the knowledge to be developed also has to be geared to the decision. To get the right science, it is necessary to ask the right questions. But identifying the questions to be addressed in a scientific analysis is not a straightforward task. For example, when assessing risk associated with a Superfund site containing hundreds of substances, or the ecological effects of introducing a new genetically engineered organism into the natural environment, there are many possible harms that could be analyzed, and it is not at all obvious which ones should be chosen. Consequently, the problems selected for analysis—hazardous substances or processes, undesirable effects, and options for action—need to be determined in consultation with the decision makers and the interested and affected parties. A risk characterization will fail to be useful if the underlying analysis addresses questions and issues that are different from those of concern to the decision makers or the interested and affected parties. The key role of problem formulation in understanding risk has been strongly stated with respect to ecological risks (U.S. Environmental Protection Agency, 1992a), but its importance extends much more broadly.

Effective risk characterization depends on an iterative process with feedback. The risk decision process is a goal-directed activity, and it should be iterative. It involves a series of tasks and feedbacks that allow for learning by all participants. Risk characterization, that is, the task of synthesis, depends on preceding tasks in the process, such as problem formulation and information gathering. It also depends on users' reactions to previous risk characterizations and decisions, which may lead to redefinition of a problem and change the needs for analysis and synthesis.

Risk characterization depends on an analytic-deliberative process. Understanding a risk depends on the interplay of two processes. *Analysis* includes various ways of reasoning and drawing conclusions by systematically applying theories and methods from natural science, social science, engineering, decision science, logic, mathematics, and law. *Deliberation* includes the methods by which people build understanding or reach consensus through discussion, reflection, persuasion, and other forms of communication—processes that allow for interaction across different groups of experts and between experts and others. Both analysis and deliberation are essential, and they interact within each of the tasks leading up to risk characterization: deliberation frames analysis, and analysis informs deliberation.

Effective risk characterization depends on the appropriate representation, involvement, or participation of the interested and affected parties. Successful risk characterization depends on input from three kinds of actors: public officials or other designated decision makers; analytic experts, such as natural and social scientists; and the interested and affected parties to the decision. Scientists, of course, are often employed by government agencies or interested parties. Public officials and some scientists are usually included in the process that leads to risk characterization, but the interested and affected parties are sometimes overlooked. Their inclusion is critical to ensure that all relevant information is included, that it is synthesized in a way that addresses the parties' concerns, and that those who may be affected by a risk decision are sufficiently well informed and involved to participate meaningfully in the decision. The interested and affected parties have a right to influence which questions should be the subject of analysis and can contribute both to developing information and to the deliberative parts of the process. The most appropriate types of inclusion will depend on the particular risk decision and may vary at different points in the process leading to a risk characterization. Sometimes, the situation may require direct participation and involvement by interested and affected individuals or groups; at other times, representation by surrogates may be most appropriate. Identifying the interested and affected parties and obtaining their appropriate involvement are important at all steps of the process that informs risk decisions.

The appropriate level of effort for a risk characterization is situation specific. The above discussion may seem to suggest that risk characterization should *always* be based on extensive public participation at every step, lengthy deliberation, detailed analysis of a great variety of possible adverse outcomes, and the like. Such is not our intention. Although we believe analysis, deliberation, and participation have too often been inappropriately restricted in processes leading to risk characterization, extensive and expensive efforts in these directions are only occasionally warranted. Judgment is critical in determining the amount, content, and timing of analysis, deliberation, and participation that are appropriate for supporting a particular risk characterization. Good judgment results from two things: careful diagnosis of the decision situation to arrive at preliminary judgments on these matters and openness to reconsidering those judgments during the process. The procedures that govern risk characterization should leave enough flexibility for the process to be expanded or simplified to suit the needs of the decision.

Our conception of risk characterization may perhaps be best seen by reference to Figures 1-1 and 1-2. In both conceptions, risk characterization involves synthesizing or summarizing information; the difference lies in who is involved in producing the characterization, what is synthesized, and how. In Figure 1-1, risk characterization is an activity conducted by experts in risk analysis that synthesizes or summarizes the results of analytical work by the same or similar experts. As this view of risk characterization is often implemented, expert judgment and statistical techniques are the most important methods used to synthesize information. Risk characterization is conceived as the last step in a process of information gathering and interpretation. It makes sense of available information, but does not affect what information has become available.

In Figure 1-2 risk characterization is conducted by a more diverse group of participants that may include, depending on the needs of the situation, not only analytic experts (labeled "natural and social scientists"), but also public officials and interested and affected parties. (The categories overlap: scientists, for example, may work for government, interested parties, or independently.) The information that goes into a risk characterization is determined by a similarly diverse group and is developed to meet their concerns, as well as the needs of the decision. Thus, risk characterization is not only the end of an analytic process, but also an important shaper of that process. The needs of risk characterization help formulate the problem for scientific analysis and influence the ways information is generated and interpreted. These relationships are represented by some of the various feedback loops in Figure 1-2. Expert judgment and statistical analysis are used to synthesize information, but this step also relies on deliberative methods that allow information to be considered from multiple perspectives. For risk characterization to meet the needs of a decision, it is important for each step of the process prior to synthesis to integrate analysis and deliberation and to involve, as appropriate, scientists, public officials, and interested and affected parties.

Figure 1-2 represents the interplay of analysis and deliberation and of these various participants throughout the process leading to a risk characterization and, beyond that, in decision making and implementation. Arrows indicate both the presence of feedbacks and the major direction of the process, which moves toward decision and action. The figure shows public officials, scientists, and interested and affected parties as participating throughout the process, which represents the default presumption. The figure presents more detail about the processes preceding a decision than those following it because it is the former processes that are our main concern. What happens during and after a decision is more complex than shown, but those details are not as directly important to the success of risk characterization.

Our definition of risk characterization will seem overly broad to some readers who think of risk characterization as simply the summary of available scientific information about risk. This narrower definition is widely used and familiar from previous National Research Council (1983) work. We found this definition wanting because it suggests that a risk characterization (and by inference the understanding of the people who use and comprehend it) is acceptable if it adequately reflects and represents existing scientific information. We have written a broader definition that highlights the fact that risk characterizations that meet this test can and do still fail when, for instance, the underlying analysis fails to address the questions that the users of the characterization see as relevant, when the characterization fails to reflect important perspectives and concerns, or when the process inappropriately restricts participation.

Our definition retains the sense of risk characterization as a synthesis or summary, but it offers a broad conception of what it should synthesize, for whom, and how it should be developed. The definition makes clear that even though risk characterization does not include all the activities represented in the arrow at the left of Figure 1-2, its success depends on the quality of all of these prior activities.

The need for a broader concept of risk characterization derives, we believe, from a shift in the roles of risk analysis and characterization in public policy over the past two decades. Before the 1970s, much of the deliberation that drove federal risk decisions occurred in Congress; by the time agency officials entered the process, legislation had already formulated the problems, defined the decision processes, and identified the adverse outcomes that would trigger regulatory action. Regulatory agencies such as the Food and Drug Administration and the Environmental Protection Agency had the task of determining facts, such as whether a chemical was a carcinogen, and their substantive decisions were supposed to follow more or less automatically from findings of fact (e.g., Interagency Regulatory Liaison Group, 1979; Rodricks, 1988; Paustenbach, 1989; Albert, 1994). When the concept of risk characterization was developed in the late 1970s and early 1980s, almost all regulatory experience with risk had been with decisions of this type. Over the past two decades, however, agencies have increasingly been called on to conduct risk analyses with less well-specified purposes—for example, to call to legislative attention new problems that may require regulation; to assess loosely formulated risk problems, such as those involving risks to ecosystems; and to address local issues, such as those of hazardous wastes, where legislation has not specified how agencies should arrive at decisions or which outcomes they should consider. A narrow concept of risk characterization and a linear view of the risk decision process may have been adequate when an organization was dealing with only a small part of the process; now, when public agencies are routinely responsible for much more of the process, a broader view is necessary.

The Risk Assessment-Risk Management Distinction

The traditional view of the risk decision process makes a sharp distinction between two functions, risk assessment (understanding) and risk management (action); see Figure 1-1. Risk assessment is usually defined as the scientific analysis and characterization of adverse effects of environmental hazards. It may include both quantitative and qualitative descriptors, but it often excludes the analysis of perceived risk, risk comparisons, and analysis of the social and economic effects of regulatory decisions (e.g., National Research Council, 1983:18). Risk assessment is often presumed to be free of value judgments, with some important exceptions, such as choices about whether and to what extent to include worst-case assumptions in risk assessments, a choice that may be made differently depending on whether the assessment is being conducted to determine regulatory priorities or priorities for testing (National Research Council, 1983:40). Risk management refers to the activities of identifying and evaluating alternative regulatory options and selecting among them. Risk managers are supposed to deal with broad social, economic, ethical, and political issues in choosing from among a set of decision options by using the results of the risk assessment and their understanding of the other issues. Making tradeoffs, which may be called risk-benefit, cost-benefit, or risk-risk evaluations, is part of risk management.

The conceptual distinction between risk assessment (understanding) and risk management (action) remains useful for various important purposes, such as insulating scientific activity from political pressure and maintaining the analytic distinction between the magnitude of a risk and the cost of coping with it. For the purposes of improving decision-relevant understanding of risk and making that understanding more widely accepted, however, a rigid distinction of this sort does not provide the most helpful conceptual framework.[4] The reason, in brief, is that the analytical activities generally considered to constitute risk assessment are not sufficient by themselves to provide the needed understanding. Much of this volume elaborates on how those activities must be shaped and complemented by deliberation in order to yield useful risk characterizations.

Our judgment partly reflects several developments since 1983. Risk characterizations have been needed for a much wider range of policy questions, including many in which the nature of the problem and the identity of the available choices is not at all obvious; articulate and scientifically informed public opposition to risk decisions has revealed gaps in many risk analyses; experiences with risk communication have demonstrated that official summaries of risk are often incomprehensible, confusing, or irrelevant to many of the affected parties; and public trust in many of the organizations that conduct risk assessments has declined. These developments underline the limitations of an approach to informing risk decisions that presumes that it is sufficient to get the science right: that the sound way to build understanding of risks is to apply methods from epidemiology, toxicology, statistics, and a small number of other scientific specialties. We believe that acceptance of too strict a separation between risk assessment and risk management has contributed to an unworkably narrow view of risk characterization.

Careful studies of the risk decision process have increasingly acknowledged the limitations of a strict separation. They were recognized first in the Red Book itself (National Research Council, 1983), which pointed to the need to iterate between risk assessment and risk management so that assessment could incorporate analytical assumptions that may need to be different for functions such as initial screening and the evaluation of regulatory options: it was noted that "a single risk assessment method may not be sufficient" (p. 40) and that the choice of appropriate assumptions required interaction between the assessment and management functions. As this study further noted (p. 142):

---

[4]We therefore generally avoid the terms risk assessment and risk management in this book although we do use the former term in discussing government agency functions or products that are normally identified by that name.

Separation of the risk assessment function from an agency's regulatory activities is likely to inhibit the interaction between assessors and regulators that is necessary for the proper interpretation of risk estimates and the evaluation of risk management options. Separation can lead to disjunction between assessment and regulatory agendas and cause delays in regulatory proceedings.

The Red Book recognized that "interpretation of risk estimates" involves an important element of judgment because of gaps in data and theoretical understandings, and this theme was elaborated much further in the more recent National Research Council (1994a) report, *Science and Judgment in Risk Assessment*. Methods of risk analysis can make only a limited contribution to improving such judgments. Progress can be made, however, by strengthening the processes, only some of which are analytical in nature, that are used for informing risk decisions.

Our framework emphasizes a series of tasks that support risk characterization and help inform risk decisions and the two processes by which these tasks are performed: analysis and deliberation. It emphasizes that there is a role for both scientific method and for appropriately broad-based deliberation in each of the tasks. It makes explicit that although good analysis is essential, it is not the only way to increase understanding among participants in risk decisions; indeed, relying on analysis alone is detrimental to the enterprise.

This framework implies that those responsible for risk decisions should look at each task differently. Before determining what information to gather, and well before considering how to summarize it for participants in a decision, they should ask several diagnostic questions, such as: What are the decisions that can or need to be made? Which outcomes for individuals, society, or the environment are of concern? Who are the potential participants—public officials, scientific and technical experts, and interested and affected parties—in the decisions? What information would be needed to address the questions or problems as identified by the participants and to satisfy them that their concerns are being given adequate consideration? Who should be involved in answering these questions so that the answers are acceptable to the participants? The answers to such questions will determine how to structure the process for informing the particular decision to be made.

Structure of the Book

The rest of this book substantiates our framework for risk characterization and sets out some of its implications for practice in government agencies and other organizations. Chapter 2 summarizes evidence that provides much of the rationale for our framework, particularly evidence of the many important judgments embedded in each step leading to risk characterization. Chapters 3, 4, and 5 focus on the analytic-deliberative process. Chapter 3 defines deliberation, explains the need for appropriately broad-based deliberation for risk characterization, and discusses some principles for organizing effective deliberation. Chapter 4 discusses the general principles and purposes of analysis in the context of risk characterization, a subject that has been elaborated in greater detail elsewhere, and focuses particularly on two analytical issues: simplifying the understanding of risk by combining many of its dimensions into one and characterizing uncertainty. Chapter 5 addresses the challenge of finding an appropriate balance of analysis and deliberation within each of the major steps of the process leading to a risk characterization.

Chapters 6 and 7 deal with implementing risk characterization. Chapter 6 discusses the implementation of our framework. It addresses the issue of practicality and the problem of matching the analytic-deliberative process to the decision. It especially emphasizes the diagnostic effort that is required at the beginning of the process. It also addresses the problem of building the capability to implement the framework. Chapter 7 presents a set of principles for implementing the process and approach to risk characterization that we advocate. Appendix A details some of the risk decision cases that are referred to briefly throughout the text, and Appendix B briefly discusses some common approaches to deliberation and public participation, noting the research literatures on them.

2

Judgment in the
Risk Decision Process

A risk characterization is part of a process that begins with the formulation of a problem—the likelihood of a harm—and ends with a decision. A risk characterization cannot make up for deficiencies in other parts of the process; inversely, if the other parts of the process are done well, it is far more likely that the risk characterization will be both clear and useful.

Some of the analytical difficulties affecting risk characterization are well known, such as the difficulty of determining an appropriate mathematical model for extrapolating from animal toxicological data to assess the health consequences of human exposures and of comparing best estimates of different risks when their uncertainty distributions differ in shape or in variability (see, e.g., Finkel, 1990; National Research Council, 1994a). In this chapter we focus on other difficulties, often overlooked in the extensive literature on risk analysis, that are equally important for understanding and coping with risk. Many of these difficulties result from judgments made at each step of the process that can undermine the quality of risk characterization and, if they are unacceptable to some of the interested and affected parties, become lightning rods for conflict. Such difficulties tend to arise when the knowledge and perspectives of these parties were not adequately incorporated into the process that led to the judgments. Many of the difficulties can be prevented or reduced if the process is recognized from the start to require both analysis and deliberation and if it is organized to ensure that the judgments are informed by appropriate deliberations.

We consider in detail the steps of problem formulation, selecting options and outcomes to consider, and information gathering, as well as synthesis, usually the major focus of risk characterization. We discuss process design at the end of the chapter. We document the variety of judgments made during each of these steps and some of the ways these judgments can undermine understanding of risks and contribute to mistrust and public conflict about risk decisions. The chapter concludes with a strategy for avoiding these outcomes by designing the analytic-deliberative process so as to inform the key judgments with the knowledge and perspectives of the range of decision participants.

Certainly, many risk characterizations and risk decision processes have been appropriate for the decision at hand. However, as we note in Chapter 1, some high-profile, controversial risk characterizations have suffered from deficiencies, and sometimes the damage to decision making has been significant. The deficiencies also threaten some lower-profile risk characterizations.

PROBLEM FORMULATION

Perhaps the most basic difficulty with risk characterization is that the people who will or should participate in the risk decision process frequently have divergent perspectives on the decision at hand. Differences of perspective cause problems because efforts to inform decisions necessarily proceed from some implicit formulation of the problem: a risk characterization that deals selectively with only one perspective on a problem will be inadequate for those with significantly different perspectives.

The Concept of Risk

Judgments pervade any understanding of risk (National Research Council, 1994a). Some writers even question the idea that risk should be conceptualized as a quantifiable physical reality (e.g., Douglas and Wildavsky, 1982; Funtowicz and Ravetz, 1992; Krimsky and Golding, 1992; Otway, 1992; Pidgeon et al., 1992; Slovic, 1992; Watson, 1981; Wynne, 1992). They argue that the concept of risk helps people interpret and cope with the dangers and uncertainties of life, including but not limited to the prospect of physical harm, and that the concept is shaped by human minds and cultures. That is, there are many different kinds and qualities of dangers and many potentially useful ways of making sense of them, and even though many of these are measurable in principle, it is judgments and values that determine which ones are defined in terms of risk and actually subjected to measurement.

The multidimensionality of risk and the many ways it can be viewed help explain why risk characterizations sometimes lack authority for some of the interested and affected parties to a decision, even when the characterizations are supported by high-quality analysis. Individuals and groups that do not share the judgments and assumptions about the problem formulation that underlie a risk characterization may well see the information it provides as invalid, illegitimate, or not pertinent. They may see the characterization as flawed because the underlying risk analysis is based on controversial assumptions (often implicit) about which perspectives are legitimate, which solutions are reasonable, and which types of information are useful or relevant (Vaughan and Seifert, 1992).

The history of risk analysis is filled with instances in which analysis, at least to some of the parties, seemed to beg the question. One such case is the risk analysis for the Yucca Mountain nuclear waste repository site, mentioned in Chapter 1. Billions of dollars were spent on assessing the quantitative and calculable risks associated with permanent disposal at one site, when many people believe it would have been more productive to assess the risks of temporary storage while engaging in a more thorough debate on the merits of a permanent solution. Another example was the comparison of coal and nuclear power generation in the 1970s that did not consider slowing the growth of energy demand as one approach to finding sufficient electric generating capacity to meet national needs.

Even the apparently straightforward act of defining the hazardous pollutant to be characterized can embed important assumptions about the nature of the problem. Should one consider narrow classes of compounds such as dioxins, or broad classes such as the thousands of organochlorines and assess chlorine as the relevant environmental risk? Such choices, though they should be well informed by toxicology and other relevant science, involve important acts of judgment that shape risk characterization and even decision making (Fischhoff et al., 1981; O'Brien, 1995). In the chlorine example, the definition of the hazard is highly consequential for the chemical industry and for proponents of pollution prevention; a risk characterization based solely on *either* formulation might be unsatisfactory to one of the interested and affected parties.

Missing Considerations

Three considerations that are often missing from the formulation of risk problems have led to disputes about the subsequent risk characterizations: fairness, prevention (of pollution or risk), and rights. Acknowledging these concerns may lead to different (usually broader) problem formulations than those that emerge from the ordinary routines of government agencies. The way such concerns are or are not addressed can directly affect choices about the options and outcomes to consider in characterizing risk.

Fairness

For some interested and affected parties in risk decisions, managing environmental risks has become a question of fairness, moral responsibility, and distributional equity (Beach, 1990; Bullard and Wright, 1992; Lawless, 1977; Nelkin, 1989; Sandman, Weinstein, and Klotz, 1987; Vaughan and Seifert, 1992). An example from Chester, Pennsylvania, shows how fairness issues can arise in risk characterization. Chester is an industrial city with a declining population (now about 40,000) consisting largely of low-income African Americans. It has become the site of numerous hazardous facilities, including two oil refineries, a trash incinerator, some Superfund sites, and an autoclave facility for infectious materials. When a proposal arose to site a soil decontamination plant in Chester, the Pennsylvania Environmental Protection Agency proposed to do a risk assessment of the plant project, examining its likely emissions and projecting the incremental health risks to the local population on the basis of models of exposures and dose-response relationships. But community representatives raised several other issues, one of which was the claim that adding a new hazardous facility was unfair in a city where residents were already bearing more than their share of toxic exposures.

The city's questions led the regional office of the U.S. Environmental Protection Agency (EPA) to agree to conduct an analysis based on a different problem formulation: a cumulative risk assessment that would characterize pollution in Chester generally and identify the areas of highest risks. Such a risk analysis would focus serious scientific attention on matters that are not considered in an incremental-risk analysis, including the interactive effects of the hazards present in the city, the health effects of the new hazard on a population whose health status may be compromised by other exposures, and the comparison of overall pollution risks in the exposed population with those in more affluent communities nearby. These issues would not have been addressed in the original state-proposed analysis. The city's position was that any risk characterization that ignored these issues would be incomplete and inadequate in terms of providing the information needed to make a major decision about public health (personal communication, Gregory Schirm, 1994).

Some fairness concerns, described in terms of "environmental justice" for minority and low-income populations, were given prominence by Presidential Executive Order 12898, issued in February 1994. The Executive Order recognized that federal agencies' risk analyses had not previously made equity issues a routine part of the problem definition and directed them to do so. Effective implementation of the order would make the analysis of some aspects of fairness and equity an essential input into risk characterization.

Prevention

Current debates about preventing pollution and risk also show clearly how problem formulation shapes risk characterization and the entire risk decision process. One proponent of pollution prevention has criticized standard practice in risk analysis for asking the question, "Which environmental problems can we ignore?" (because their risks are negligible), rather than the question, "How can we avoid exposures to hazards?" (O'Brien, 1995). While both formulations omit the key question of cost, the latter question invites consideration of a much wider range of possible policy options (especially, pollution prevention). For many interested and affected parties, a risk characterization that does not address prevention could never provide the information they need to accept a risk decision, no matter how well the narrower problem is analyzed or characterized.

The controversy over control of the Mediterranean fruit fly, a major pest to the $2 billion fruit and vegetable industry in California, illustrates, among other things, how judgments about which aspects of risk to characterize can obscure or highlight debates over pollution prevention. Much of the Medfly debate and the associated risk analysis and characterization in the early 1990s focused on estimates of risks to the health of residents who might be exposed to the malathion spray. The debate centered on dose-response questions and on the appropriateness of assumptions about the behavior—and therefore the exposure—of residents who are warned to stay indoors during an imminent spraying but who might not do so. It is based on the formulation that the Medfly problem in California comes from flies that have recently entered the state on infested fruit and that spraying where flies have been seen will eradicate the problem (California Department of Food and Agriculture, 1994).

Some critics argue that this formulation is incorrect because the flies are now established as breeding populations in California and cannot be eradicated by a spraying program (Carey, 1991; 1994). They argue further that the state has a vested interest in the isolated infestation assumption because under it spraying might easily and cheaply convince potential importers of California produce that it is fly free. If, however, the Medfly is an established pest, malathion will not perform as claimed, and biological pest control would become an alternative option worth considering. Biological control would also avoid the projected human exposures to malathion spray. The formulation of the problem as one of keeping an established pest population under control changes the questions for risk analysis: the focus shifts to the risks of biological control approaches and the comparative effectiveness of chemical and biological Medfly control; it makes pollution prevention, in the form of nonchemical means of pest control, a key policy option. If the Medfly problem has been incorrectly formulated, much of the risk analysis and characterization has focused on an incomplete set of options, and the needs of decision participants for understanding have been poorly served.

Rights

Risk characterizations have also become controversial when they pay little attention to issues of the rights of individuals or groups to control their own lives. An example is the continuing debate about the fluoridation of public water supplies. Advocates declare that this technique is a safe and cost-effective method of improving people's dental health. Opponents—in addition to questioning the certainty of the scientific estimates and raising the issues of increased risk of bone fractures in the elderly and fluoridosis in those with poor kidney function—speak about individual rights and the undesirability of a public health policy that eliminates individual choice in regard to exposure to a chemical agent that has both benefits and risks (Lawless, 1977; Martin, 1989).

Another example is the contentious debate about environmental "unfunded mandates," federal decisions that require states and localities to spend their money on certain environmental and health projects, which consequently reduces funds available for other projects. Many interested and affected parties claim that they have rights to clean air and water and that the federal government should protect these rights. Other parties, chiefly state and local officials, complain that federal decisions that compel their action, even if based on sound risk analyses, abrogate the right of localities to use their funds to reduce risks in the most cost-effective ways. While federal officials may have sponsored risk analyses to focus on reducing the risks of environmental chemicals, a state public health department or a mayor faces a different problem: a choice between reducing risks to citizens from chemical residues in water or from birth defects, traffic accidents, or violent crime. A risk characterization for water pollution alone may be quite beside the point for a local official confronting such a choice.

SELECTION OF OPTIONS AND OUTCOMES

Problem formulation has practical implications for other steps in the risk decision process. Among the most important is the way it shapes choices about which options to consider and which possible adverse outcomes to analyze—choices that are critical to the success of risk characterization.[1] For a risk characterization to meet the needs of participants in a decision, it must consider the range of plausible decision options. The parties to a decision may not agree on which options are worth considering, but a risk characterization that does not consider an option that one of the participants views as promising is likely to be seen as biased and inadequate. The controversies over the Medfly and unfunded mandates are examples. When problem definitions truncate the list of options too severely, risk characterizations can be doomed to controversy long before they are undertaken.

Organizations responsible for risk characterizations should make efforts to identify the range of decision options that experts and the spectrum of interested and affected parties consider viable. Generating an adequate list of options may be difficult. It demands familiarity with the context of the decision, knowledge about the scientific and technical aspects of the possible risks, and, sometimes, creativity and imagination. Also, it often demands that organizations listen to the interested and affected parties. Although identifying the range of options is challenging, it is a key to successful risk characterization.

Considering a sufficiently broad range of possible harms or losses is equally important to risk characterization. Typically, analysis focuses on only a few adverse outcomes—such as cancer or birth defects in human beings, loss of a species, or elimination of a habitat—that are judged to be the most serious of the possible harms. The analysis is sometimes further restricted to the effects of exposure to a particular agent (a substance or process), through a particular medium (e.g., air, water, or food intake). Agencies may narrow the list of outcomes because they have narrow responsibilities under law, because funds for analysis are limited, because of political pressures, or for other reasons.

Consideration of only a few possible outcomes is usually justified on the assumption that if a decision protects adequately against the selected outcomes, it will also protect against the others, because environmental and health hazards are strongly correlated. This assumption becomes increasingly suspect as the range of outcomes of concern expands from overt human health risks to effects on the immune system and related systems (e.g., allergenicity); behavioral effects; psychological effects, such

---

[1]Selection of outcomes has sometimes been treated as a stage within problem formulation: for example, the U.S. Environmental Protection Agency (1992a:12) refers to outcomes as "ecologically based endpoints". We prefer to make a sharper conceptual distinction between the two tasks.

as anxiety and depression; ecological effects; and social, economic, and ethical impacts. Some participants may doubt that protection against one of these adverse outcomes is tantamount to protection against all. Some of the nonmeasured outcomes may be more salient to some parties, and they may argue that these other adverse outcomes should be subject to analysis and characterization. Demands for such expansion of the outcomes to be included can be especially meritorious when data are sparse on certain outcomes, but the risks have nevertheless been dismissed on the assumption that they are small enough to ignore.

An adequate risk characterization must address all the outcomes or consequences of a hazardous situation that are reasonably important to the relevant public officials and to the interested and affected parties to the decision. Agencies should tailor their analyses to the decision to be made, addressing the potential adverse outcomes most significant for that decision.

Ecological Effects

One important class of nonhealth outcomes is harm to nonhuman organisms and ecosystems. The EPA has taken the lead in developing a conceptual framework for conducting ecological risk assessment (U.S. Environmental Protection Agency, 1992a, 1992b, 1992c, 1992f, 1992g, 1993h) and is preparing guidelines for this activity. Analysis is difficult because the effects may fall on individual animals or plants, on local populations of a certain species, on ecosystems (thus affecting many species), or on the survival of endangered species. At larger scales, effects on the distribution of ecological communities across the landscape are central to regional-scale ecosystem management (Grumbine, 1994; Harwell et al., in press). There may be important ecological outcomes to consider and characterize at each of these hierarchical levels of ecological systems (Harwell et al., 1990).

Ecological risk analysis requires an understanding of how the affected ecosystem functions. There are numerous interrelationships among taxa, across responses, and across organizational levels. In addition, some of the most important effects may be indirect, operating through several interrelationships. Many of these effects are inadequately understood, difficult to measure, or laden with uncertainty (National Research Council, 1993a). Some ecologists even dispute whether the concept of ecological risk (or its inverse, ecological health) is useful for policy analysis (e.g., Lackey, 1994, 1995). None of these scientific difficulties of estimation, however, negate the importance for policy decisions of considering ecological outcomes. Interested and affected parties may want to take account of ecological effects even if the level of scientific understanding of them is poor. Qualitative assessments of relative ecological risks can provide useful insights for environmental decision making (Harwell et al., 1992). A critical need is to develop appropriate tools for assessing the value of ecological systems, including both economic and noneconomic (e.g., intrinsic) values.

Economic and Social Effects

Economic consequences are sometimes inextricable from the other aspects of a risk. We are not referring here to the well-recognized economic costs of regulating risks, but to the economic costs of the hazards themselves. Many risk characterizations do not consider the full range of adverse economic outcomes, even though they are important for decision purposes and amenable to scientific analysis. A 1987 example from Brazil is illustrative. When two men seeking scrap metal pried open a metal capsule containing 100 grams of cesium 137, subsequent exposures to neighbors and family resulted in 4 deaths and 50 cases of radioactive contamination that required medical treatment. Described in this sense, the accident appears to have been of local significance, without major national or international impact. But that depiction fails to capture the $20 million dollars in cleanup costs and the subsequent 50 percent drop in the wholesale value of agricultural products from the Brazilian state in which the accident occurred. Sales of manufactured goods were also affected, despite the lack of plausible contamination (Freudenburg, 1988). The incident illustrates that losses in terms of human health may not be the only adverse outcomes of a hazardous situation that are worth characterizing.

One important class of economic effects are those a given hazard has on nearby property values (Greenberg, 1995; Gregory, Flynn, and Slovic, 1995). A relationship between changes in property values and proximity to hazardous waste sites has been repeatedly demonstrated (Greenberg and Hughes, 1992, 1993; Schulze et al., 1994; Skaburskis, 1989). For families living near hazardous facilities, property value losses are sometimes a significant factor that they want estimated and considered in a decision process. Such information can provide guidance for making decisions about the costs of risk remediation plans (McClelland et al., 1990). Another kind of economic effect is the cost of insurance premiums and emergency preparedness that flow directly from the possibility of an adverse event (Freudenburg, 1988). These costs are borne by the potentially affected population regardless of whether they actually suffer from the adverse event.

Risk characterizations typically do not address social effects, perhaps because they are considered outside the purview of formal risk analysis.

Yet they are legitimate objects for risk characterization because participants in decisions need to understand them to make informed choices, and many social effects are amenable to systematic analysis. Social effects that may need to be considered in a risk decision include neighborhood disruption and issues of social equity and stigma (Gregory, Flynn, and Slovic, 1995). Some risk decisions can significantly alter a community's character. Neighbors often express fears that a hazardous facility will be destructive to the community (Zeiss and Atwater, 1991). It is not uncommon to see these kinds of concerns taken seriously in negotiations about sitings, but they are not usually treated in conventional risk characterizations.

In part, what is at stake is community control (Elliot, 1984; Zeiss and Atwater, 1991). People are more willing to tolerate a risk if they feel they have some control over the exposure (Slovic, 1987). By the same token, removing control from a community has social costs, even if the community accepts a risk. For example, cleaning up a contaminated site near a residential community is a disruptive process. Several studies have documented the tension between residents who want total removal of contamination (usually younger families with children) and those who want minimal disturbance and cost (usually older couples living on fixed incomes who have owned their homes for many years and have no children living at home). The latter group sometimes opposes clean-ups that would disrupt their lives (e.g., remove their gardens, demand temporary evacuation, or involve high costs), while the former group finds the status quo more disruptive (Claus, 1995; Fessenden-Raden et al., 1987; Levine, 1982). Risk analyses could, but rarely do, explicitly consider the effects of such kinds of neighborhood disruption. If the analyses implicitly set this potential loss equal to zero, affected parties may find the risk characterization unsatisfactory.

Many affected parties in risk decisions expect the government to endeavor to achieve some fair balance between the risks a community or an individual bears and the benefits received. Recently, this expectation has been voiced as a concern for environmental justice for minority communities (Bullard and Wright, 1992; Greenberg, 1993). There is evidence of unequal distribution of noxious facilities between communities as a function of economic and racial or ethnic differences (Bullard, 1990; Commission for Racial Justice, 1987; U.S. Environmental Protection Agency, 1992h) and of differential harm associated with these risk sources (Greenberg, 1995). However, equity concerns had rarely been considered in conventional risk assessments until Executive Order 12898, in 1994. To the extent that this directive is implemented, agency risk characterizations will begin to provide information that will inform public deliberation on equity issues.[2]

People sometimes hold negative associations for things, places, organizations, or people they connect to risks (Slovic, 1993b). Such stigma can have a tangible economic impact: in 1989, concerns about the use of Alar on apples led to nationwide decline in apple sales of over $100 million (about 10%) after risk assessments were publicized that linked the substance to cancer in children (Rosen 1990). Researchers have also identified the potential economic effects from stigma associated with the proposal to construct a high-level radioactive waste facility in Yucca Mountain, Nevada (Slovic, Layman, et al., 1991). For many siting decisions, the effects of stigma cannot be reduced by engineering and design alone, but it may be possible to address them through compensation or insurance (Fort, Rosenman, and Budd, 1993). Although these effects or potential losses from stigma are difficult to quantify or compensate (Gregory, Flynn, and Slovic, 1995), they are nevertheless important to consider.

Effects on Future Generations

Many risk decisions may impose risks on future generations that require a different kind of consideration from risks to people living today. The high-level nuclear waste disposal facility planned for Yucca Mountain, Nevada, is a striking example: releases of radioactive material from this facility could cause harm thousands of years in the future. Such situations present two questions for risk analysis: How can one be certain that the risks to future generations are known? How can one represent the interests of future generations in a current risk decision process?

The difficulty of the first question is illustrated well by the Yucca Mountain controversy. As described in Chapter 1, a fundamental assumption of U.S. and international policy on radioactive waste disposal has been that safe, permanent disposal was the strategy most likely to reduce the risks to future generations. But a 1993 technical review committee set up by the state of Nevada—one of the interested and affected parties—questioned even that most basic assumption. Arguing that no one today can predict what human beings might be able, or motivated, to do at the Yucca Mountain site over the next 10,000 years, the Technical Review Committee (1993:14) concluded that rather than protecting future

---

[2]Various formal analytical techniques exist for informing discussions about distributional equity: all involve controversial techniques for valuing human lives (e.g., Zeckhauser, 1975; Anderson, 1988; Leigh, 1989; Ellis, 1993).

generations, entombment leaves them in charge of dangerous waste, while "making it as difficult as the state of our technology permits" for them to do anything about it if future knowledge or social conditions require such action. That report suggests, among other things, that people today cannot assess the risks to future generations without carefully considering possible social changes as well as the operation of physical and biological processes over the long term.

The second question (the interests of future generations) is sometimes addressed by using economic techniques of time discounting (see, e.g., Viscusi and Moore, 1989; Cropper, Aydede, and Portney, 1994). However, this practice reduces the significance of risks that lie more than a generation or two in the future almost to zero. This technique is unacceptable to many people. Another way to address the interests of future generations is to bypass explicit analysis on the assumption that living persons can act as proxies for future generations. But this strategy is also vulnerable to criticism because it assumes that people in the future would support decisions made by people today and accept the processes by which those decisions were reached (see Shrader-Frechette, 1993b). Although intergenerational equity is difficult to resolve by formal or quantitative analysis, it nevertheless raises important issues for risk characterization.

Ripple Effects

Some hazards have "ripple effects" (Slovic, 1987)—effects that extend far beyond their direct harms and that can impose very large costs. Companies in an industry may be affected by a mishap, regardless of whether they caused it; even industries unrelated to the mishap may be affected. The case of the radioactive materials accident in Brazil, described above, is an example. Another is the 1979 nuclear power accident at Three Mile Island, Pennsylvania, which killed no one, but had extensive ripple effects. It devastated the utility that operated the plant; resulted in greatly increased costs for regulating, constructing and operating nuclear power plants; and led to "reduced operation of reactors worldwide, greater public opposition to nuclear power, and reliance on more expensive energy sources" (Slovic, 1987:201; see also Evans and Hope, 1984; Heising and George, 1986). Researchers are working to understand the mechanisms that produce these ripple effects, which Kasperson and his colleagues (Kasperson et al., 1988; Kasperson, Golding, and Tuler, 1992) call the "social amplification of risk" (see also Mazur, 1981, 1984; Kunreuther and Linnerooth, 1982; Slovic, Lichtenstein, and Fischhoff, 1984).

Effects on Democracy, Governance, and Ethical Beliefs

Some of the actions that may be taken in response to risks can have widespread reverberations throughout society. An example that relates to governance is the federal imposition of unfunded mandates for environmental protection on state and local governments, noted above. The affected governments complain that federal risk decisions decrease their power and that of their constituents to control their lives, and they want this risk considered before policy decisions are made. Another example, noted in Chapter 1, is the possibility that risk decisions and the way they are made may undermine public trust in the organizations making the decisions. This effect may not only make it more difficult to implement a decision, but may lead some individuals to distrust all information provided by an organization, withdraw from the decision, or express frustration in destructive ways. The legitimacy of government may, in this sense, be one of the things at risk, although this risk is rarely characterized formally.

Decisions about risks may also violate deeply held values or ethical beliefs of affected parties. For example, during the early years of the cold war, decisions were made by the U.S. Department of Defense and the Atomic Energy Commission (a predecessor of the U.S. Department of Energy) to conduct radiation experiments without the informed consent of those exposed in some of the experiments, on the grounds of national security. When these decisions became public, there was widespread outrage. In other cases, policies may threaten what some people see as the intrinsic value of natural phenomena, such as the survival of species and the maintenance of ecosystems, and violate their belief that humanity has no right to interfere with these natural phenomena. The possibility that a risk decision will violate such ideas of what is morally right is rarely given explicit attention in risk characterization.

Conclusion

We do not suggest that all conceivable options for action or possible adverse outcomes can or should be the subject of detailed analysis in every risk decision process. In most instances, such detailed analyses would be unnecessary, not to mention the demands they would put on analysts and on scarce resources. We recognize that government agencies and other organizations must make decisions and that some of these will inevitably be opposed by some of the interested and affected parties. We emphasize, however, that the options and outcomes that risk analysts and managers traditionally choose for analysis may not be the only ones that are necessary to analyze and characterize for the decision at hand. These choices deserve careful and explicit consideration before analysis begins. (We discuss strategies for making such choices in Chapter 6.)

INFORMATION GATHERING AND INTERPRETATION

After the risk problem, the options for action, and the important outcomes have been defined, analysts, together with public officials and interested and affected parties, gather and interpret information. This task, like the others, involves judgments that can create problems in risk characterization. This section discusses two key types of judgments that color information gathering and interpretation and can affect the success of a risk characterization: choosing a risk measure and making simplifying assumptions.

Choosing a Risk Measure

Even the apparently simple task of choosing a risk measure for a well-defined outcome such as human fatalities can be surprisingly complex and judgmental. The list below shows a few of the many different ways that risks of death have been measured:

deaths per million people in the population
deaths per million people within x miles of the source of exposure
deaths per unit of concentration
deaths per facility
deaths per ton of toxic substance released
deaths per ton of toxic substance absorbed by people
deaths per ton of chemical produced
deaths per million dollars of product produced
loss of life expectancy associated with exposure to the hazard

The choice of a measure can make a big difference in a risk analysis, especially when one risk is compared with another. It can also make a big difference in whether interested and affected parties see the analysis as legitimate and informative.

An example from coal mining demonstrates how the choice of one measure or another can make a technology look either more or less risky (Crouch and Wilson, 1982). Between 1950 and 1970 coal mines became much less risky in terms of deaths from accidents per ton of coal, but they became marginally riskier in terms of deaths from accidents per employee: see Figures 2-1 and 2-2. Which measure one thinks more useful for informing decisions depends on one's point of view (Crouch and Wilson, 1982:12-13):

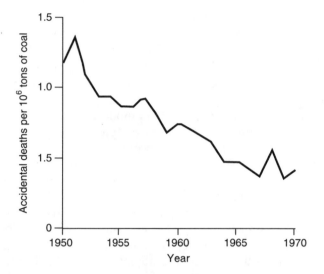

FIGURE 2-1. Accidental deaths per million tons of coal mined in the United States, 1950-1970. SOURCE: Crouch and Wilson (1982:12). Used with permission.

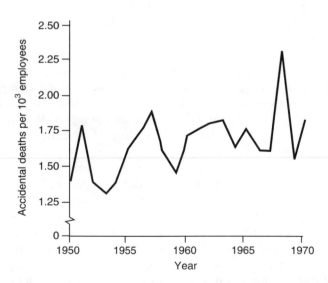

FIGURE 2-2. Accidental deaths per thousand coal mine employees in the United States, 1950-1970. SOURCE: Crouch and Wilson (1982:13). Used with permission.

> From a national point of view, given that a certain amount of coal has to be obtained, deaths per million tons of coal is the more appropriate measure of risk, whereas from a labor leader's point of view, deaths per thousand persons employed may be more relevant.

A risk analysis that presented either measure of fatalities, by itself, might well be seen by some participants as misleading.

Every way of summarizing deaths embodies its own set of values (National Research Council, 1989). For example, reduction in life expectancy treats deaths of young people as more important than deaths of older people, who have less life expectancy to lose. Simply counting fatalities treats deaths of the old and the young as equivalent; it also treats as equivalent deaths that come immediately after mishaps and deaths that follow painful and debilitating disease. Also in the case of delayed illness and death, a simple count of adverse outcomes places no value on what happens to exposed people who may spend years living in daily fear of illness, even if they ultimately do not die from the hazard.

Using number of deaths as the summary indicator of risk implies that it is as important to prevent deaths of people who engage in an activity by choice as it is to prevent deaths of those who bear its effects unwillingly. Thus, the death of a motorcyclist in an accident is given the same weight as the death of the pedestrian hit by the motorcycle. It also implies that it is as important to protect people who have been benefiting from a risky activity or technology as it is to protect those who get no benefit from it. One can easily imagine a range of arguments to justify different kinds of unequal weightings for different kinds of deaths, but to arrive at any selection requires a judgment about which deaths one considers most undesirable. To treat all deaths as equal also involves a judgment. In sum, even so simple and fundamental a choice as how to measure fatalities is value laden. It can present a dilemma in which no single summary measure, no matter how carefully the underlying analysis is done, can satisfy the expectations of all the participants in a risk decision process. Other methods may be needed to allow the parties' various perspectives to be addressed.

Needless to say, the difficulties of choosing a measure expand when the adverse outcomes are less precisely defined. Measures of morbidity, for example, raise questions of judgment about which measures appropriately aggregate different types of morbidity. Should morbidity be measured in terms of the value of lost work time? If so, is it appropriate to value at zero the health of people who do not work in paying jobs? Should severity of incapacitation be measured, and if so, how? Should long illnesses count the same as multiple short illnesses with the same total duration?

Measuring risks to ecosystems present additional judgments because of uncertainties about such matters as which ecological changes constitute threats and whether measurement should focus on biotic populations, species, habitats, or other levels of analysis. Measuring economic and social risks requires still other judgments. Measuring each type of outcome presents its particular set of judgments, and each judgment embeds values.

Making Simplifying Assumptions

Risk analysis requires making simplifying assumptions when information is incomplete or difficult to gather by regularly used methods. For example, a toxicologist's quantitative estimate of a chemical's carcinogenic risk may be based on theoretical models and assumptions that are partly subjective and depend on judgment. There are many such assumptions: about how toxic substances cause cancer and how the body resists toxins; about the shape of the dose-response function; and that the cause of the cancer can be modeled as resulting from a single chemical without taking into account other unknown or identified causative factors. Nonscientists' risk models and assumptions likewise rest on simplifying assumptions about the physical and social worlds (e.g., Kempton, 1991; Bostrom, Fischhoff, and Morgan, 1992). Although these models are rarely as consistent or mathematical as scientists' models, they may be no more subjective or dependent on judgment.

Simplifying assumptions generate especially serious problems when some of the assumptions are unreasonable in the face of information available to people outside the analytical process. For example, sometimes decision makers understandably rely on generalizations, and direct risk analysts to do the same, even though local conditions lie outside the range of applicability of the particular generalizations. The contamination of British pasturelands in Cumbria from the Chernobyl nuclear plant accident presents an example of assumptions that were unreasonable because they misrepresented local conditions (Wynne, 1989). British scientists based their advice on the assumption that radioactive cesium would quickly become immobilized in soils and so would not pose a long-term threat to the sheep feeding on local grass. Apparently, however, that assumption was based on the response of the clay mineral soils of southern England. The high-organic matter, acidic soils of Cumbria did not immobilize cesium as expected. It remained available for root uptake into the grass and found its way into the bodies of the sheep. Being unaware of the local soil conditions, or unaware that cesium behaves differently in different kinds of soils, public officials made a decision and gave advice that turned out to be wrong: they assured farmers the exposure to their lambs would last only a few weeks when in fact the problem lasted much longer.

Simplifying assumptions may also misrepresent local habits and customs that affect the incidence or magnitude of risks. Indeed, the success of exposure assessments relies on being able to accurately model the behavior of individuals. In epidemiological studies, considerable effort is expended to document patterns of behavior so that risks can be calculated for different groups of exposed individuals. Failure to consider the habits and customs of populations in sufficient detail may undermine simplifying assumptions and be directly responsible for events that cause loss. For example, part of the debate about the health effects of malathion spraying for Medfly control concerned whether local residents—some of whom did not understand English—would respond appropriately to broadcast warnings to stay indoors during sprayings.

Risk characterizations and the resulting decisions can fail because they include incorrect assumptions about geographical, economic, structural, organizational, and other conditions that may constrain the way those at risk respond to a hazard. For example, in the Cumbrian sheep farming areas after the Chernobyl accident, British officials set up a system for keeping contaminated lambs off the market. They demanded that farmers apply for permission to sell lambs 5 days before the actual sale, not recognizing the farmers' needs to act spontaneously as conditions change (health of lambs, market volume, location of market at which to sell, condition of pasture, etc.). Officials also advised farmers to keep the lambs out of the valleys to minimize radiation exposure, but the farmers found such advice preposterous because they could not control their flocks' movements in the unfenced fields (Wynne, 1989). The risk estimates erred by assuming management strategies that farmers could not reasonably be expected to adopt.

Risk characterizations often implicitly (and inaccurately) assume that individuals will do as instructed and that organizations will function as routines or regulations specify. In estimating migrant farm workers' exposures to pesticides, for example, risk analysts may be directed to assume that pesticides are applied as required by regulations and that workers wear the prescribed protective gear. This assumption may be unreasonable: Inspection of the working conditions where migrants are employed suggests a much different pattern of behavior, both of the growers and the workers. One result of assuming that rules are followed is, in this instance, a serious underestimate of actual exposures, in part because large numbers of migrants do not regularly use self-protective measures (Vaughan, 1993a, 1993b). Individuals may fail to follow instructions because of inability to read or understand them, failure to make sense of the language of risk estimation, lack of motivation to comply, various pressures for noncompliance, a belief that their actions will not really reduce risk, or other reasons. Such factors alter risks of many kinds, including the risks of pesticides to farm workers (Vaughan, 1993a) and various conventional health risks (Ell and Nishimoto, 1989; Peterson and Stunkard, 1989; Vaughan, 1995).

Risk analyses and characterizations sometimes make unreasonable assumptions about so-called human factors, such as breakdowns in the interaction between equipment and its operators; unanticipated interventions of "outsiders" (from disgruntled former employees to uninformed legislators); "organizational factors," such as failures of commitment to controlling risks, bureaucratic attenuation of information flows, diffusion of responsibility, coordination problems among subunits, and the low status of safety and maintenance units in many organizations; the atrophy of vigilance over time, both in individuals and organizationally; and a skewed distribution of organizational resources (e.g., Perrow, 1984; Shrivastava, 1987; Freudenburg, 1992; Clarke, 1993; Clarke and Short, 1993). Unrealistic assumptions that such phenomena are unimportant can be easily recognized by individuals with long experience observing the relevant behaviors, who have a kind of expertise the professional risk analysts may lack.

Risk characterizations often assume that organizations will function according to plan in a crisis. The validity of this assumption is critical for accurately understanding many risks, including those from nuclear power plant failures, various kinds of industrial and shipping accidents, and air traffic accidents. These situations present workers, work teams, and organizations with the generic problem of shifting from a routine mode of operation to a crisis mode without losing effectiveness. For individuals, performance depends on the ability to function well under stress; with fatigue or sleep disruption; and in the context of environmental stressors such as heat, noise, and vibration. For individuals and work groups, performance depends on the ability to switch tasks effectively, to manage task priorities, to retain skills and routines from past training, and to meet challenges of leadership and coordination. Individuals and groups do often meet all these challenges, and crisis responses are often quite effective, but research does not support the baseline assumption that task groups will reliably function as planned (see National Research Council, 1993b). Risk characterizations based on such an assumption, especially if they do not consider the past record of an organization in managing the particular risk or other relevant evidence on individual and organizational performance, are likely to be misleading and to be criticized as inadequate by people who are well acquainted with the organization or the kinds of behavioral changes a particular crisis demands.

Risk characterizations may be based on unrealistic assumptions about

how well an organization will stay vigilant against low-probability catas-
trophes. For example, for the first several years after the Alaska pipeline
opened, the Alyeska Pipeline Service Company maintained an emergency
response team and escorted each tanker out of Valdez harbor with a tug
as precautionary measures. Both practices were abandoned after several
years without tanker accidents, so neither was in place in March 1979
when the *Exxon Valdez* grounded on leaving the harbor, spilling 260,000
barrels of oil into Prince William Sound (Clarke, 1993).

Alyeska has also been criticized for systematically disregarding information suggesting that a catastrophe was likely; it had drafted and probably believed in unreasonably optimistic contingency plans (Clarke, 1993).
Organizational tendencies, such as to disregard warnings about possible
dangers, have also been implicated in the accident at Three Mile Island
(President's Commission on the Accident at Three Mile Island, 1979) and
the explosion of the space shuttle *Challenger* (Paté-Cornell and Fischbeck,
1993; Presidential Commission on the Space Shuttle *Challenger* Accident,
1986; Vaughan, 1990). Some analysts see such organizational responses
as predictable, especially given production pressures and the tendency
for bad news to be filtered out as it passes up the organizational chain of
command (Freudenburg, 1992; Clarke, 1993). The state of an
organization's emergency preparedness is relevant for risk characterization, and it is more appropriate to estimate it, when possible, on the basis
of information than on general assumptions about organizational behavior.

SYNTHESIS

Synthesizing information is a well-recognized difficulty that affects
risk characterization, and it will remain so even if all the other steps in the
process better address the issues discussed above. This section considers
four major sources of difficulties in synthesizing information: summarization; the multidimensional nature of risk; the meaning of risk estimates;
and communication.

Summarization

The fundamental challenge in synthesis, from a technical standpoint,
is to produce an unbiased summary of existing knowledge. Even a single
piece of scientific evidence can often be summarized in various ways,
equally correct and truthful, that convey strikingly different understandings or meanings to audiences. One example (discussed above) is fatality
estimates from the Chernobyl accident. U.S. analysts summarized the
risk in terms of the absolute numbers of excess cancer deaths predicted

from a linear no-threshold model—about 50,000, which seemed high; the Soviet government summarized the information from the same model as a percentage increase in deaths among the millions of people who had been exposed—an increase of about one-quarter of 1 percent, which seemed low (Smith, 1986).

Numerous research studies have demonstrated that different (but logically equivalent) ways of summarizing the same risk information can lead to different understandings and different preferences for decisions. One dramatic example comes from a study that asked people to imagine that they had lung cancer and had to choose between two therapies, surgery or radiation (McNeil et al., 1982). The two therapies were described in some detail. Then, some subjects were presented with the cumulative probabilities of surviving for varying lengths of time after the treatment, while other subjects received the same cumulative probabilities, but framed in terms of dying rather than surviving (see the table below). For example, one group was told that 68 percent of those having surgery will have survived after 1 year, and the other group was told that 32 percent will have died. As the table shows, framing the statistics in terms of survival lowered the percentage of subjects choosing radiation therapy over surgery from 44 percent to 18 percent (McNeil et al., 1982):

| | Mode of Summarization | | | |
| | Mortality Rates | | Survival Rates | |
	Surgery	Radiation	Surgery	Radiation
Treatment	10%	0%	90%	100%
After 1 year	32%	23%	68%	77%
After 5 years	66%	78%	34%	22%
Subjects choosing radiation therapy	44%		18%	

The effect was as strong when the subjects were physicians as when they were lay people.

Such systematic differences in preferences that depend on the way information is summarized or "framed" can be explained by the prospect theory of Kahneman and Tversky (1979), which has been applied to the question of presenting risk information for policy purposes (see, e.g., Cole and Whithey, 1981; Gregory, Lichtenstein, and MacGregor, 1993; Heimer, 1988; Stern, 1991). According to prospect theory, outcomes of a decision are evaluated as gains or losses from some reference point—usually the status quo. The psychological impact of these gains and losses follows a "value function," such as that shown in Figure 2-3. According to this function, the impacts of gains and losses are nonlinearly related to their magnitudes. That is, gaining $200 gives less than twice the value that is obtained from gaining $100. Moreover, losses have more impact than comparable gains: a $200 loss hurts more than a $200 gain pleases.

Prospect theory leads to some dramatic illustrations of the effects of subtle variations in problem framing, including the above example of lung cancer treatment. Another example comes from a public health problem given to separate groups of respondents (Tversky and Kahneman, 1981):

> Problem 1. Imagine that the United States is preparing for the outbreak of an unusual Asian disease, which is expected to kill 600 people. Two alternative programs to combat the disease have been proposed. Assume that the consequences of the programs are as follows: If Program A is adopted, 200 people will be saved. If Program B is adopted, there is a one-third probability that 600 people will be saved and a two-thirds probability that no people will be saved. Which of the two programs would you favor?

> Problem 2. (Same introduction as in Problem 1.) If Program C is adopted, 400 people will die. If Program D is adopted, there is a

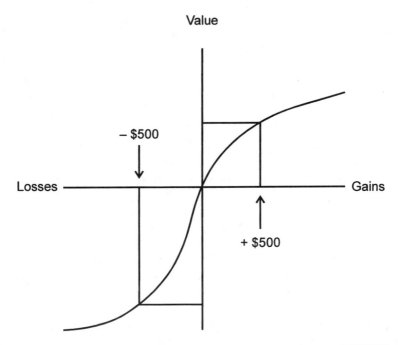

FIGURE 2-3. A hypothetical value function in prospect theory. SOURCE: Kahneman and Tversky (1979). Used with permission of the Econometric Society.

one-third probability that nobody will die and a two-thirds probability that 600 people will die. Which of the two programs would you favor?

The two problems are formally identical—Programs A and C are the same, as are Programs B and D—but the preferences tend to be quite different. In one study, 72 percent of the respondents given Problem 1 chose Program A over Program B, while 78 percent of those given Problem 2 chose Program D (which is formally equivalent to Program B) over Program C (formally equivalent to Program A). This reversal of preference was predicted by prospect theory on the basis of the concept of a reference point and the nonlinearity of the value function: Program A is chosen over Program B because people see little additional gain from saving 600 lives (which is uncertain), and Program D is chosen over Program C because people see little extra loss from 600 deaths (which might not even occur) in comparison with 400 deaths that are certain.

The policy implications of characterizing risks in terms of potential gains or potential losses can be important. Vaughan and Seifert (1992) note that in a typical decision situation, any choice or policy strategy involves the acceptance of some nonzero level of risk, as well as potential economic or other gains. Such situations easily lend themselves to alternative conceptualizations, which may highlight either what is to be gained or what is to be lost by a particular course of action. In California, for example, the debate about eradicating the Medfly through aerial spraying of malathion over populated areas brought into focus two contrasting formulations of the policy question. Some groups initially framed the problem as a consideration of what options would minimize the chances of a loss of millions of dollars for California's agricultural industry. Others framed the question in terms of whether the relative gains associated with aerial spraying were sufficient to justify accepting any additional risks to human health (Roe, 1989). Similarly, debates about the costs of decreasing workers' exposure to occupational hazards often feature two contrasting positions: one evaluates protective action in terms of maximizing the numbers of lives saved per dollar; the other evaluates the action in terms of the number of lives that could be lost if additional safety provisions are not implemented (Hilgartner, 1985). In the past, several major public controversies over technological and environmental issues have been marked by contrasting frames that differed by describing policy options either in terms of potential gains or potential losses (e.g., Brunner, 1991; Heimer, 1988; Lawless, 1977).

Prospect theory also implies that decision makers who differ in their views of the status quo will choose different policy options because they begin the decision task from different reference points. For example, a

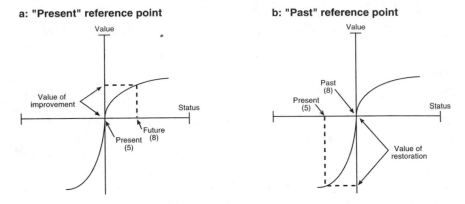

FIGURE 2-4. The value of an improvement in prospect theory. SOURCE: Gregory, Lichtenstein, and MacGregor, 1993. Used with permission.

study asked people to evaluate the desirability of improving the water quality in Oregon's Willamette River from its present state of Grade 5 on a 10-point scale to Grade 8 (Gregory, Lichtenstein, and MacGregor, 1993). In one condition, respondents were told that the river once had quality equivalent to Grade 8; thus, the improvement would represent a restoration of lost quality. A second group of respondents were told that the change represented an improvement from the current level of 5 to a level of 8. These two framing conditions, which differed in terms of how they characterized the status quo, are illustrated in Figure 2-4, within the context of the value function from prospect theory. Because of the asymmetry of the value function (steeper for losses than for gains), prospect theory predicts that the improvement from 5 to 8 will be more attractive when framed as restoring lost quality (2-4b) than when framed as improving the present quality (2-4a). Indeed, the study found that the desirability of water of Grade 8 was greater for people who believed this quality signified a restoration of lost water quality.[3]

These examples demonstrate that every way of presenting risk information is a "frame" that can shape the judgments of the participants in a risk decision. The same information can be presented as lives saved or lives lost, mortality rates or survival rates, restoring lost (or "natural")

---

[3]This finding might also be interpreted in other ways. Respondents may have placed inherent value on nature and therefore believed that water should be kept at its "natural" quality. They may also have interpreted the information that the water quality was once Grade 8 as evidence that this was an achievable goal.

water quality or improving present water quality, and so forth. Neither frame is right or wrong—they are just different. There is no scientific way to determine that one summary is more correct, or less biased, than another when both accurately reflect the data. Thus, the problem of generating a single unbiased summary of information about risk to meet the needs of participants in risk decisions has no purely technical solution. Any decision about how to synthesize risk information involves judgments of considerable practical importance. Because subtle differences in how risk is summarized can have marked effects on understanding, those responsible for synthesis may have considerable ability to influence perceptions and behaviors. This possibility creates procedural and ethical problems that experts and public officials must recognize and address in their efforts to characterize risks (MacLean, 1995). In Chapters 3, 4, and 5 we discuss ways to address these problems by combining analysis with deliberation to arrive at a publicly acceptable, decision-oriented synthesis of available risk information.

The Multidimensional Nature of Risk

Risk characterizations often focus on a single outcome, most often human fatalities, but as discussed above, even a single outcome has multiple attributes. Furthermore, many risk decisions involve multiple outcomes, so that there are at least several attributes and kinds of information to synthesize. The general problem is how to characterize what is known about a risk when there is no clear way to combine its many attributes into a single scale or metric.

Over the past several decades, research on how people understand, think about, and react to risk has shown that judgments of risk can be described in terms of numerous characteristics or dimensions. Figure 2-5 presents a spatial display of hazards within a perceptual space derived from individual judgments by people who were not experts in risk analysis. The factors in this space reflect the degree to which a risk is perceived to be known or understood (vertical dimension) and the degree to which it evokes perceptions of dread, uncontrollability, and catastrophe (horizontal dimension).

People's response to risk is closely related to the position of a hazard within this space. In particular, the further to the right a hazard appears, the higher its perceived risk, the more people want to see its current risks reduced, and the more they want to see strict regulation to reduce the risk (Slovic, 1987). In contrast, specialists in risk analysis tend to understand risk in ways not closely related to these dimensions or the characteristics that underlie them. Instead, they tend to see riskiness as synonymous, especially for policy purposes, with expected annual mortality, consistent

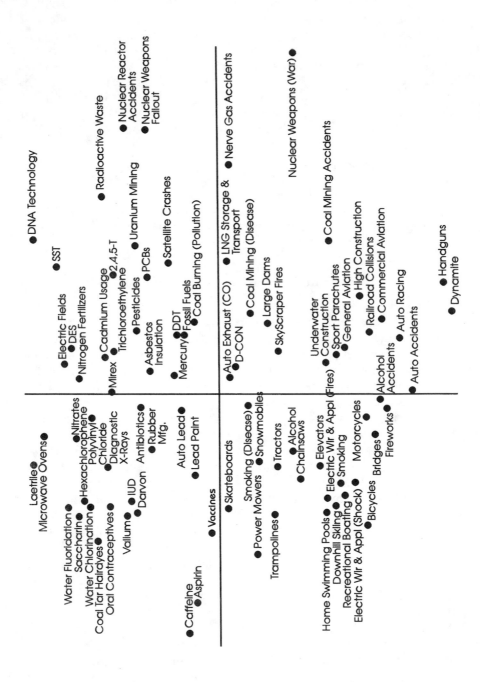

- DNA Technology
- Nuclear Reactor Accidents
- Nuclear Weapons Fallout
- Radioactive Waste
- Nuclear Weapons Accidents
- Nerve Gas Accidents
- Nuclear Weapons (War)
- SST
- Electric Fields
- DES
- Nitrogen Fertilizers
- Cadmium Usage
- 2,4,5-T
- Trichloroethylene
- Uranium Mining
- PCBs
- Pesticides
- Satellite Crashes
- Coal Burning (Pollution)
- LNG Storage & Transport
- Coal Mining (Disease)
- Large Dams
- Coal Mining Accidents
- SkyScraper Fires
- High Construction
- Underwater Construction
- Sport Parachutes
- General Aviation
- Commercial Aviation
- Railroad Collisions
- Auto Racing
- Handguns
- Dynamite
- Mirex
- Asbestos Insulation
- Mercury
- DDT
- Fossil Fuels
- Auto Exhaust (CO)
- D-CON
- Alcohol Accidents
- Auto Accidents
- Laetrile
- Microwave Ovens
- Nitrates
- Hexachlorophene
- Polyvinyl Chloride
- Saccharin
- Water Fluoridation
- Water Chlorination
- Coal Tar Hairdyes
- Oral Contraceptives
- Diagnostic X-Rays
- Valium
- IUD
- Darvon
- Antibiotics
- Rubber Mfg.
- Auto Lead
- Lead Paint
- Caffeine
- Aspirin
- Vaccines
- Skateboards
- Smoking (Disease)
- Snowmobiles
- Power Mowers
- Tractors
- Alcohol
- Chainsaws
- Trampolines
- Elevators
- Electric Wlr & Appl (Fires)
- Smoking
- Motorcycles
- Home Swimming Pools
- Downhill Skiing
- Recreational Boating
- Electric Wlr & Appl (Shock)
- Bicycles
- Bridges
- Fireworks

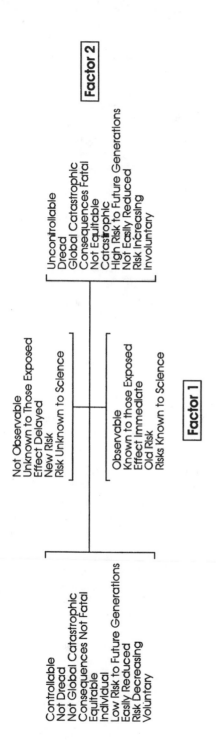

FIGURE 2-5. A schematic representation of understandings of risk among nonexperts. The locations of 81 hazards in Factors 1 and 2 are derived from the interrelationships among 15 risk characteristics. Each factor is made up of a number of characteristics, as shown in the lower part of the figure. SOURCE: Slovic (1987:282); copyright 1987 American Association for the Advancement of Science. Reprinted with permission.

with the ways that risks tend to be characterized in quantitative risk assessments (Slovic, Fischhoff, and Lichtenstein, 1979; The Royal Society Study Group, 1992; Lindell and Malmfors, 1994).

Conflicts over "risk" may reflect differences between specialists in risk analysis and others on their definitions of the concept. In this light, it is not surprising that citations of statistics about "actual risks" often do little to change most people's attitudes and perceptions. Nonspecialists factor complex, qualitative considerations into their estimates of risk, including judgments about uncertainty, dread, catastrophic potential, controllability, equity, and risk to future generations.

The legitimate, value-laden issues that underlie these multiple dimensions of risk need to be considered in risk policy decisions (Fischhoff, Watson, and Hope, 1984). For example: Is risk from cancer (a dread disease) worse than risk from auto accidents (not so dreaded)? Is a risk imposed on a child more serious than a risk accepted voluntarily by an adult? Are the deaths of 50 passengers in separate automobile accidents equivalent to the deaths of 50 passengers in one airplane crash? Is the risk from an industrial emission worse if the facility is located in a neighborhood that has a number of other hazardous facilities nearby? The difficult questions multiply when outcomes other than human health and safety are also considered. As noted in an earlier study (National Research Council, 1989:51):

> Technological choices sometimes involve weighing the value of a river vista, a small town style of living, a holy place, or the survival of an endangered species, in addition to human health, against probable benefits. Such matters are ultimately matters of values.

The fact that hazards differ dramatically in their attributes or characteristics helps explain why certain technologies or activities, such as nuclear power, evoke much more intense public opposition than others, such as motorcycle riding, that cause many more fatalities. The implications of "risk perception" for synthesis have been well described in a previous study (National Research Council, 1989:52):

> Those quantitative risk analyses that convert all types of human health hazard to a single metric carry an implicit value-based assumption that all deaths or shortenings of life are equivalent in terms of the importance of avoiding them. The risk perception research shows not only that the equating of risks with different attributes is value laden, but also that the values adopted by this practice differ from those held by most people. For most people, deaths and injuries are not equal—some kinds or circumstances of harm are more to be avoided than others. One need not conclude that quantitative risk analysis should weight the risks to conform to majority values. But the research does suggest that it is presumptuous for technical experts to act as if they know, without

careful thought and analysis, the proper weights to use to equate one type of hazard with another. *When lay and expert values differ, reducing different kinds of hazard to a common metric (such as number of fatalities per year) and presenting comparisons only on that metric have great potential to produce misunderstanding and conflict and to engender mistrust of expertise.*

This analysis is still pertinent with reference both to techniques for characterizing relative risks in terms of expected deaths and to techniques that compare hazards on a common monetary metric, such as contingent valuation methods or various forms of cost-benefit analysis.

A number of risk analysts have sought technical solutions to the problem of taking qualitative aspects of risk into account. Generally, they have proposed broadening risk analysis to incorporate one or more of the various characteristics identified in studies of perceived risk: for example, distinguishing between voluntary and involuntary activities in assessing risk-benefit balances (Starr, 1969); giving proportionally more weight to large accidents than to numerous small accidents that cause the same amount of damage or number of deaths (Wilson, 1975; Griesemeyer and Okrent, 1981); and adjusting risk estimates to take into account the importance of various risk-perception characteristics (Rowe, 1977; Litai, Lanning, and Rasmussen, 1983). None of these proposals has yet been developed to the point of application in actual risk assessments.

A related approach has successfully integrated several of the dimensions of risk in a formal way and done so in real applications. This approach has been developed by two Swiss analysts to aid decisions about the safety of ammunition storage depots and transportation systems, including the design of a high-speed railway in Germany (Bohnenblust and Schneider, 1984). The method characterizes the risk reduction in terms of cost-effectiveness, and it attributes more value to reducing risks that are involuntary, poorly understood, potentially catastrophic, and hard to control. Although this Swiss model has been applied to a number of important decision problems, there is a need to align the model more closely with recent research that has been done on social, cultural, and psychological factors (see, e.g., Krimsky and Golding, 1992). This might be accomplished, at least in principle, by refining quantitative approaches to risk analysis (see Chapter 4); it might also be achieved qualitatively through deliberative processes (see Chapters 3 and 5).

Another way to incorporate some of these dimensions into risk characterizations is to apply multiattribute utility analysis (e.g., von Winterfeldt and Edwards, 1986; Fischhoff, Watson, and Hope, 1984; Gregory, Lichtenstein, and Slovic, 1993; Brody and Rosen, 1994; see also Appendix A). In this technique, individuals identify value dimensions, attributes, or outcomes that are important to them, assign relative weights to them, and evaluate the outcomes identified by risk analysis (or a set of policy options) on each attribute. Multiattribute utility analysis allows individuals to compare options that yield different packages of risks and benefits. It allows for evaluations to be explicitly subjective: individuals can, for example, assign a numerical value to a quality such as "incidental encounters with neighbors." But it does not, by itself, solve the problem of providing risk estimates for populations because there is no acceptable formula for aggregating individuals' evaluations. Should each individual's evaluation have equal weight, or should those who might bear the risk have a greater weight? This problem of estimating risks for whole groups when the risks are of various kinds pushes the limits of analysis. (In Chapter 4 we discuss the strategy of using analytical techniques to reduce the dimensionality of risk. In Chapter 5 we discuss the alternative strategy of combining deliberative processes with analysis to help the participants in decisions develop working understandings of multiattribute risks.)

The Meaning of Risk Estimates

Separable from the technical questions about how best to estimate the risk of a particular agent with respect to a particular outcome is the question of what the risk estimate means, or should mean, to participants in risk decisions. Risk characterizations often fail because they attribute meaning to scientific estimates in ways that mislead participants in the risk decision process or that are incomprehensible to them. This section discusses several such sources of failure: in the treatment of uncertainty, in inferences about which populations will be affected, and in inferences about how a risk estimate should be interpreted in light of other risks to which a population is exposed.

Uncertainty

Risk characterizations often give misleading information about uncertainty in several ways. They may give the impression of more scientific certainty or unanimity than exists (or of more uncertainty or dissension). They may suggest that uncertainty is a matter of measurement when in fact it is a matter of disagreement about whether a particular theory applies or differences in judgment about how to infer something that is unknown from something that is known. And they may give the impression that certain risks do not exist when in fact they have not been analyzed.

Civil engineers, public health professionals, and others often take account of uncertainty by a strategy of "conservatism." This means that they recommend decisions or actions that leave a margin for error that is intended to protect the public if the actual risk turns out to be greater than they predict. It is often argued that risk analysts should instead present their best available estimate to decision makers, along with an explicit characterization of its uncertainty, and allow the decision makers to decide explicitly how much margin of safety to allow.[4] Either approach embodies a value choice about the best way to characterize risk and protect public health and safety, and there is no scientific technique for determining which approach is preferable (National Research Council, 1994a).

There is strong agreement that risk analysts should explicitly summarize uncertainty, and there are methods for doing so (e.g., Morgan and Henrion, 1990). But despite the admonitions of officials in some government agencies (Habicht, 1992; Browner, 1995) and the recommendations of outside panels (e.g., National Research Council, 1994a), many risk characterizations still present point estimates of risk, representing these as upper-bound estimates and providing little or no analysis of the extent of overestimation. In spite of the obvious shortcomings of point estimates and the efforts to develop alternative ways of describing uncertainty— such as with probability distributions or scenario approaches—no alternative has gained widespread recognition as acceptable and practical within EPA and other regulatory agencies. It is difficult to characterize what is known about uncertainty without making the risk appear either larger or smaller than analysts believe it to be (see, e.g., Johnson and Slovic, 1995). And the difficulties are not yet yielding to analysis: the more that characterizing uncertainty is debated as an analytical problem, the more complex it appears to be (see, e.g., Finkel, 1990; National Research Council, 1994a:Chapter 9).

Characterizing uncertainty analytically puts risk analysts on the horns of a dilemma: simple characterizations are likely to give an erroneous impression of the extent of uncertainty, but more careful and elaborate characterizations may be incomprehensible to nonspecialists and so unusable by decision makers and some other participants. Like the problem of finding a single, unbiased summary of accepted scientific knowledge, the problem of summarizing uncertainty may have no technical solution. We believe, however, that a solution might be found in the *processes* that lead to a risk decision, processes that combine iterative deliberation and analysis and provide participants with enough understanding of uncertainty to appreciate where scientists agree and where they disagree. (The last section of this chapter outlines key issues in process design; Chapter 4 presents a more detailed discussion of understanding uncertainty.)

---

[4]For an illuminating exchange of views on the "conservatism" issue, see Finkel (1994) and McClellan and North (1994).

Specific Populations

The question of who is at risk is important both for decision makers and the persons whose health and safety is of concern (Vaughan and Seifert, 1992; see also Konheim, 1988; Nelkin, 1989). In presenting aggregate risk estimates to community residents who are concerned about a hazard, officials often fail to answer the important question for many interested and affected parties: "What does this mean for me or my family?" (e.g., Sharlin, 1986). When confronted with statistical risk estimates, people often seek to reframe the question in terms of personal risk (Plough and Krimsky, 1987; Siegel and Gibson, 1988)—an issue not addressed by the aggregate numbers. If a risk characterization does not address such questions, both it and the analysis behind it may be discredited by participants in the decision process. If it does address them, different ways of framing the same information can create different understandings.

Such failings often arise when a risk characterization assumes that estimates of risks to one population are sufficient to answer risk questions about what may be a different population. For example, the 1989 controversy over the use of Alar on apples centered on the risks to children, who drink more apple juice than adults, but the standard was set on the assumption that an adult male weighing 70 kilograms was an adequate surrogate for everyone (Jasanoff, 1987). The risk reduction expected from vaccination programs is usually presented for the entire population although rural children, who live far from treatment centers, may not receive as much benefit as other children because vaccination programs may not reach them. Migrant farm workers and their families may be inadequately protected against workplace risks because standards are based on exposure under very different working conditions (Vaughan, 1993a).

To adequately summarize risks to some populations may require behavioral analyses as well as the traditional analyses of exposure and sensitivity. For example, different groups of farm workers who are exposed to pesticides vary in their ability to understand warning materials and in their propensity to take self-protective action when given the opportunity (Vaughan, 1993a). Knowledge about reading ability and the psychological factors underlying self-protective behaviors are not usually incorporated in risk characterizations although they can obviously affect both the risks to exposed individuals and the effectiveness of options to reduce those risks. Thus, a risk characterization that fails to carefully consider which population(s) the estimate is for may be inadequate to inform decision making.

Multiple Exposures

Risk analyses for multiple exposures are often based on the assumption that risk from genotoxic carcinogens are additive and that non-cancer risks are not, unless the agents operate by similar mechanisms. The degree to which this assumption holds is subject to debate because of the limited data available to address it. It is even less clear how well the assumption may apply for combinations of biological, chemical, and physical risks to an ecosystem. EPA has considered multiple chemical ecological risks, but only at certain sites and within narrow bands. Although the agency also recognizes physical hazards and those that arise from management practices (U.S. Environmental Protection Agency, 1992a), it has not considered their possible interactions. For humans, evidence of serious drug interactions suggests that even with chemical hazards there are some instances in which the assumption of additivity may be questionable. Increasing concerns about synergistic effects warrant careful consideration of how to address them in risk characterizations.

A related issue is the treatment of past exposures or the past health conditions of some of the population at risk. This issue may arise as one of health (e.g., the possibility that past exposures have synergistic effects with present ones) or of equity. As the Chester, Pennsylvania, case suggests, residents of an industrial community who believe that they have already had more than their share of exposure to chemical risks may demand on equity grounds that past exposures be considered as part of the risk characterization.

Communication

The success of a risk characterization depends on its effective delivery to the participants in a risk decision. Typically, not all participants will understand a risk message in the same manner. Analogies are often used to make risk summaries more understandable, but analogies are usually very specific and sometimes depend on culture, status, age, gender, and other characteristics for their interpretation. If the manner in which the risk message will be interpreted by different groups or participants is not considered, uneven risk protection across groups could result (Vaughan, 1993a, 1995). Another even more fundamental problem is that of comprehension. Non-English-speaking people obviously get no benefit from a risk characterization in English. Messages prepared in written form will be ignored by people who cannot read or who are used to receiving information in other forms.

The history of interaction between an organization that is presenting a risk characterization and the interested and affected parties can be another source of communication problems (e.g., Krimsky and Plough, 1988). A party that has had unsatisfactory experiences with that organization or that issue may simply be unreceptive to new information from that source. For example, in a decision-making process to permit experimental land application of sewage sludge on farmland, an elaborate public involvement process collapsed partly because the same community had been involved in a landfill siting controversy just one year earlier. A widespread belief that the community was targeted as a "dumping ground" overpowered any positive reaction to the public involvement plans. People in the community organized against the new proposal partly because of ill feelings toward the state regulatory agency (Renn et al., 1991).

CONCLUSION: THE IMPORTANCE OF PROCESS DESIGN

This chapter surveys the variety of judgments made—sometimes, without careful consideration—in the course of analyzing and characterizing risk that can become lightning rods for controversy. They become problematic when they conflict with the judgments of some of the interested and affected parties to a decision, so that the resulting risk characterization does not address these parties' needs. The best way to prevent such problems, we believe, is not to call all such judgments into question in every decision process. Doing this would make risk analysis and characterization inordinately complex and resource intensive. We believe the best preventive is to devise analytic-deliberative processes that will pay appropriate attention to the judgments involved in problem formulation and the other tasks, inform these judgments with the best available knowledge and the perspectives of the spectrum of decision participants, and thus guide risk characterizations toward addressing the needs of the decision.

When understandings of risk depend on potentially controversial judgments, it seems prudent to involve those who are likely to be affected by the decisions that rely on those judgments. If, for instance, there are many scientifically defensible ways of counting deaths and if the choice has serious implications for the concerns of some of the interested and affected parties, it makes sense to involve those parties in selecting the measures of death that will be used to characterize risk. Organizations responsible for characterizing risk should anticipate the value-based judgments that are likely to become contentious in the context of a particular risk characterization and consider putting them on the agenda for the analytic-deliberative process.

We emphasize strongly that improved risk characterization based on a better designed process will not eliminate conflict about risk. The best it can hope to do is to eliminate or reduce those conflicts that are based on misunderstandings, mistrust, miscommunication, inadvertent neglect of a point of view, and the like. It might be said that although good practice does not predictably lessen conflict, bad practice predictably increases it.

Designing an analytic-deliberative process involves many choices. Who should be involved in the tasks that support risk characterization, beginning with problem formulation? In what ways and through what procedures should they be involved? At what points in the process should they be involved? Under what conditions should past assumptions, conclusions, or decisions be reconsidered? These choices can affect the ultimate content of a risk characterization, the ways participants in a decision understand the risks, and acceptance of the process.

Federal agency officials with a legislative mandate to protect the public against dangerous exposures to a toxic substance commonly respond to preliminary evidence of a possible hazard by directing toxicologists, epidemiologists, and other technical experts on the hazard to estimate the health risks associated with the substance. The process involves these experts, agency officials and policy makers, attendees at any required public hearings (whose ideas may or may not be given serious consideration), and any legislators and interest groups that know about the pending decision and are able to gain access to the process. This standard process often leads to objections from interested and affected parties that they have been disenfranchised, that their ideas have been ignored, that their concerns have not been taken seriously, that the risk analysis was incomplete or irrelevant, that the analyses are so complex and arcane that they cannot participate meaningfully, and so forth—in short, serious disaffection with the process and the resulting risk characterization.

Such outcomes have led many observers to recommend increased public involvement in risk decision making, better two-way communication between agencies and interested and affected parties, involvement of these parties early in the decision process, and other changes that would make risk decision making processes more broadly participatory (e.g., Kunreuther, Fitzgerald, and Aarts, 1993; Leroy and Nadler, 1993; Slovic, 1993a; National Research Council, 1994b). We agree that more complete involvement of interested and affected parties in risk characterization is often essential for improving the process. It can also be essential for arriving at sound analyses. We note here some key principles of increasing meaningful participation in risk characterization that are developed in more detail in the next several chapters:

- give explicit attention to the design of the process that informs risk decisions;

- solicit and seriously consider input from the interested and affected parties as appropriate at various points in the process leading to risk characterizations; and
- plan for iteration in the decision process, that is, for reconsidering past assumptions, conclusions, and process-related decisions on the basis of new data and changes in the decision situation.

We reiterate that risk characterization is more than a synthesis of information developed by analytical techniques. Analysis has inherent limitations in the face of the multidimensional and value-laden nature of many risk decisions. The success of risk characterization depends not only on doing and describing analysis well, but also on choosing analyses that address the needs of decision participants and on making the choice through a process that those parties trust. Organizations responsible for characterizing risks should plan to blend analysis with deliberative processes that clarify the concerns of interested and affected parties, help prevent avoidable errors, offer a balanced and nuanced understanding of the state of knowledge, and ensure adequately broad participation for a given risk decision.

3

Deliberation

Successful risk characterization depends on an analytic-deliberative process. This chapter and the next explain what we mean by the terms *deliberation* and *analysis* and how both are important in each of the tasks involved in understanding risks. Analysis and deliberation can be thought of as two complementary approaches to gaining knowledge of the world, forming understandings on the basis of knowledge, and reaching agreement among people. *Analysis* uses rigorous, replicable methods, evaluated under the agreed protocols of an expert community—such as those of disciplines in the natural, social, or decision sciences, as well as mathematics, logic, and law—to arrive at answers to factual questions. It operates on the assumption that facts can be found through an objective, that is, dispassionate and impartial, examination of phenomena. *Deliberation* is any formal or informal process for communication and for raising and collectively considering issues.

In deliberation, people confer, ponder, exchange views, consider evidence, reflect on matters of mutual interest, negotiate, and attempt to persuade each other. Deliberation includes both consensual communication processes and adversarial ones. The adjectival form of the word, *deliberate*, also implies intentionality, purpose, and a sense of having carefully thought out the consequences of actions.

Thus deliberation implies an iterative process that moves toward closure. It considers each aspect of an issue and it may revisit earlier discussion on the basis of new knowledge or insights. A good deliberative process deepens participants' understandings as details emerge and the relations among different positions become clearer. Effective deliberative processes arrive at understandings that most participants consider adequate or acceptable within the existing limits of time or effort and that all recognize to be subject to reconsideration in the future.

Participants in deliberations may have divergent interests, but deliberation need not be a tug of war among these interests. (See Mansbridge, 1983, 1990, for discussion of the distinction between deliberative and adversarial democracy.) Deliberation does not assume consensus; it brings into consideration knowledge and judgments coming from various perspectives so that participants develop understandings that are informed by other views. At its best, deliberation becomes an interactive learning process for those involved.

Deliberation captures part of the meaning of democracy (the normative rationale for participation) and contributes to making decisions more rational and legitimate (the substantive and instrumental rationales) (Fiorino, 1990). It is particularly important for building understanding and acceptance when an issue has more sides than any one participant is likely to consider without input from others. Risk, as we have outlined, is often such an issue: people with different values and interests often develop conflicting understandings of the same risk situations (e.g., Whittemore, 1983; Dietz and Rycroft, 1987; Jasanoff, 1987; Johnson and Covello, 1987; Lynn, 1986; Clarke, 1988; Dietz, Stern, and Rycroft, 1989; Dake, 1991; Kraus, Malmfors, and Slovic, 1991; Flynn, Slovic, and Mertz, 1994; Peters and Slovic, 1995.)

In the process leading to risk characterization, deliberation may involve various combinations of scientific and technical specialists, public officials, and interested and affected parties. The role of deliberation for considering conflicts of values and interests is well known and important in risk decision making. Its role in understanding risk has been relatively neglected, however. We use the term analytic-deliberative process to signify the intimate connection between analysis and deliberation and their equal importance for understanding risk.

ROLE OF DELIBERATION

Deliberative processes are important not only for democratic decision making, but also for developing the understanding required to inform decisions. In fact, deliberation has always been a crucial element in scientific progress. Scientific peer review is a form of deliberation involving an exchange of judgments about the methodological appropriateness of research methods, the strength of an author's inferences from findings, and even the likely validity of a surprising research result. Reviewers consider both scientific data and other kinds of information, such as the reputation of a researcher who reports surprising results. This is one way in which scientific communities arrive at collective understandings through a combination of analysis and deliberation. The classic statement of the point that science advances by deliberation and not just by analysis is by Kuhn (1970). Deliberation provides a way to uncover errors and deepen understanding by considering the evidence from various perspectives. Appropriately structured deliberation complements analysis by adding knowledge and perspectives that improve understanding.

In emphasizing the complementarity of analysis and deliberation and the importance of broad participation, this volume is part of an evolution from previous National Research Council studies of risk issues. Some of these have called for increased attention to matters of scientific judgment, suggesting the importance of explicit attention to deliberation. For example, the 1983 study of risk assessment emphasized "the interplay of science and policy in risk assessment" (National Research Council, 1983: 33) and the class of judgments it labeled "risk assessment policy" (p. 37). Risk assessment policy implies deliberation involving scientists and public officials as a way to guide the conduct of analysis. The 1994 report, *Science and Judgment in Risk Assessment* described in great detail a set of judgments, such as about how to analyze and present various kinds of uncertainties in risk analysis to decision makers.

Previous Research Council studies have also emphasized the need to include interested and affected parties at various points at which they are not now routinely consulted. We cite just a few such conclusions, drawn from recent studies. The *Science and Judgment* report, which reviewed EPA's risk assessment procedures for hazardous air pollutants, concluded (National Research Council, 1994a:267):

> EPA should provide a process for public review and comment with a requirement that it respond, so that outside parties can be assured that the methods used in risk assessments are scientifically justifiable.

A review of the Department of Energy's environmental remediation program for nuclear weapons sites concluded (National Research Council, 1994b:3;26):

> . . . risk assessment concerning possible future outcomes at DOE weapons-complex sites . . . must involve the public (in its many guises) in the whole process, including the planning of the process and the definition of the scope of risk assessment . . . [and] the first and probably most important step in effective risk assessment and risk management is to establish public participation that involves all the stakeholders.

A review of procedures used by the Departments of Defense and Energy and the EPA to rank hazardous waste sites for remediation recommended (National Research Council, 1994c):

> The process of developing a model (or any component of a model) [for use in ranking] should be as open as possible, involving both stakeholders and the technical community. Value preferences should be explicit in the models, and . . . [t]he process of applying the model to a given site (or to a large installation such as a military base or a DOE facility) should be similarly open, so that there is the greatest understanding of the results of the model.

These statements are significant in part because the Research Council's primary concern was to provide advice on improving analysis: despite this emphasis, the reports repeatedly noted the importance of judgment, broader public participation, and the need for public officials to listen more carefully to nonspecialists' concerns about risks. Chapter 2 shows in detail why broad participation is often necessary for sound analysis. The previous studies' conclusions point to a variety of kinds of deliberation that are desirable as part of the effort to inform decisions. This volume takes the next step, offering government agencies and other organizations more specific advice about how to plan and carry out the kinds of deliberations these previous study panels have advocated, how to integrate them with analysis, and how to use them to inform decisions.

We emphasize five points at the outset. First, deliberation is important at every step of the process that informs risk decisions.

Second, deliberation needs to be much more extensive in some decision situations and in some steps than in others. The conveners of a process leading to a risk characterization must organize deliberative processes that steer between two shoals: being so concerned with reaching closure that the process excludes important perspectives, diminishes understanding, and threatens the acceptance of decisions; and being so concerned with inclusiveness and completeness that decisions are unnecessarily delayed. Government agencies sometimes need routines for making decisions quickly, and the deliberative strategies we discuss can be quite useful in establishing these routines.

Third, there is little systematic knowledge about what works in public participation, deliberation, and the coordination of deliberation and analysis. When government agencies and other organizations have promoted or created specific deliberative processes, they have rarely reported the results of their efforts. Thus, our analysis and guidance are based largely on case material, scholarly reviews of case-based literature, and the collective experience of committee members and others who have worked in risk decision making. References to the literature on particular methods of deliberation and public participation can be found in Appendix B. We discuss coordination further in Chapter 5. We encourage organizations responsible for risk characterization to explore the possibilities for improving deliberation and to make a commitment to learn from experience.

Fourth, there are significant limitations and pitfalls associated with adopting the strategy of broadening the analytic-deliberative process. Better risk characterizations do not necessarily yield better decisions; moreover, the proposed strategy may take more time, cost more money, or play into the hands of those who would benefit from delay and might demand broader analysis or more deliberation to justify that delay. We discuss these and other problems in more detail in "Limitations and Challenges," below, and in Chapter 6.

Fifth, deliberation already occurs throughout the risk decision process, although not always self-consciously. Organizations deliberate about the best way to define and approach a risk problem before setting the tasks for risk assessment. They may also deliberate about whether to consult outsiders about setting these tasks, whom to consult, and how to design the consultative process. They may deliberate about the agenda for analysis: which options to consider; which harms to investigate (e.g., which human health effects, which ecological effects); or about other, more technical components of analysis. Scientists may deliberate about how best to summarize particular sets of findings from analysis when describing risks to high-level agency officials or to journalists, interest groups, and individual citizens. Managers may deliberate about how to coordinate work or allocate resources and funds. Scientists on expert review panels may deliberate on the appropriateness of making certain assumptions or on the validity of certain data collection designs, the proper interpretation of data, and the best way to summarize it. Individuals and groups representing public officials, scientists, or the various interested and affected parties may also participate in the process.

We see each risk decision as having a spectrum of interested and affected parties who vary in the kinds of specific knowledge they have and in their perspectives, concerns, and vested interests. *Broadly based deliberations* are those in which, in addition to the involvement of appropriate policy makers and specialists in risk analysis, participation from across this spectrum of parties is sufficiently diverse to ensure that the important, decision-relevant knowledge enters the process, that the important perspectives are considered, and that the parties' legitimate concerns about the inclusiveness and openness of the process are addressed. Such deliberation involves the participation or at least the representation of the relevant range of interests, values, and outlooks as well as the relevant range of expertise. The amount, kind, and timing of participation necessary to meet these criteria are necessarily situation dependent. For example, circumstances that provoke public skepticism will require stronger efforts at inclusion. Routine decisions may not require much attention to breadth of participation if the routines themselves were designed by an appropriate deliberative process.

Broadly based deliberation is not equivalent to public participation as that term is generally understood. One difference is the range of participants. Deliberation to improve understanding may require including not only representatives of "the public" generally, but a variety of scientists and other experts, some of whom may speak for public agencies or interested and affected parties. In a broadly based deliberation, the knowledge and perspectives of the spectrum of interested and affected parties are represented. Another difference is the venue: deliberation occurs not only in public settings, but also in the course of the ordinary activities of the agency or other organizations developing a risk characterization. A third important difference concerns timing. Public participation conventionally refers to the involvement of interested and affected parties in the decision-making task, but we are focusing on the tasks farther "upstream," those that inform decisions. Too often the deliberative process is broadened only after a risk characterization has been completed, as with "decide, announce, defend" approaches to dealing with interested and affected parties.

The common practice of eliciting comments only after most of the work of reaching a decision has been done is a cause for resentment of risk decisions. An appropriate critique of this practice is reflected in the public participation practitioners' adage: "involve the public early and often" (e.g., Kasperson, 1986). This rule of thumb leaves many crucial questions unanswered, however: Who is "the public"? Should everyone participate, or only representatives of each segment of "the public"? If the latter, how does one identify the segments? When should participation occur, and for what purposes? Are different kinds of participation appropriate for different purposes? Our answer is that the important questions do not concern whether participation or deliberation should occur, but what kind, among whom, and for what purposes.

The practitioners' adage is a good starting point. It calls for a needed change in what appears to be a default assumption of some government agencies: that interested and affected parties should be involved in the tasks leading to risk characterization only when and how this is legally required. We propose that agencies operate from an alternative presumption: that deliberation is necessary and appropriate at every step in the process. We do not advocate unlimited participation or full deliberation at every step. Rather, we advocate that agencies (and other organizations) begin by asking *how* to involve the parties in the steps leading up to risk characterization and *what* to deliberate, rather than asking *whether* to involve them. Many decisions can be better informed and their information base can be more credible if the interested and affected parties are appropriately and effectively involved in deliberation. The best way to do this will depend on the particular task, the decision context, and the party. For some routine decisions, little effort of this kind will be necessary; for others, it may be necessary to provide some parties with such resources as travel money or expert assistance so they can participate meaningfully. These points are elaborated in Chapter 5.

PURPOSES OF BROADLY BASED DELIBERATION

Involving the spectrum of interested and affected parties in deliberation can make the process leading to risk characterization more democratic, legitimate, and informative for decision participants. It has this potential in several ways: improving problem formulation, providing more knowledge, determining appropriate uses for controversial analytic techniques, clarifying views, and making decisions more acceptable.

The relationships between broadly based deliberation about risk, on one hand, and democracy and legitimacy on the other, have been elaborated in detail by others (e.g. Wynne, 1987; Fischer, 1990; Laird, 1993; Renn, Webler, and Weidemann, 1995; Sclove, 1995). These relationships provide strong justification for broadly based deliberation about risk, especially at the point of making substantive decisions about how to cope with it. This section emphasizes the role of deliberation in informing those decisions, a topic somewhat more to the point of risk characterization and one that has not been as well developed in research (but see, e.g., Dietz, 1987, 1994, and an emerging literature on participatory research on risk, e.g., Brown, 1990; Fischer, 1993; Sclove, 1995).

Broadly based deliberation can be used to frame a problem so that knowledge generated about it will be relevant to the needs and understandings of the various parties to a decision. Such deliberation may improve the quality of a solution or increase the likelihood of finding novel solutions by redefining problems (Gray, 1989). For example, the South Florida ecosystem management project (see Appendix A) used broadly based deliberation to arrive at its formulation of the problem as a choice among development plans. The participants believed that this formulation would encourage analyses that would address issues of concern to all who would be affected by development in the region.

Broadening the base for deliberation can improve the knowledge base for decisions (Ozawa, 1991). For example, as illustrated in Chapter 2, nonscientists may have critical information—such as knowledge of local conditions—that can be used to check the reasonableness of assumptions incorporated into technical analysis. Appropriate environmental monitoring may also depend on knowledge of local conditions. Broadly based deliberation can also help ensure that analysis addresses the problems that concern the interested and affected parties. The goal is to avoid studies that require many years and dollars, yet fail to advance risk characterization.

For example, as we noted above, characterizations of the risks associated with a radioactive waste repository at Yucca Mountain were not satisfactory to some of the interested and affected parties because they failed to go beyond health impacts to address other outcomes, such as the potential economic impacts—both positive and negative—for the surrounding area. The risk characterizations also failed to address many Nevadans' sense that it was unfair for their state, which has no nuclear power plants, to be singled out as the potential site for the nation's nuclear wastes. At the Hanford, Washington, nuclear weapons production site, deliberation by a broadly based working group considering remediation issues led to a call for different analyses depending on whether or not a particular area was slated for future agricultural use (Hanford Future Sites Working Group, 1992). Such broadly based deliberation can help make analysis more effective and make risk characterization more responsive to the needs of all the parties to a decision.

Broadly based deliberation can help determine appropriate uses for potentially controversial analytical techniques. For example, techniques such as cost-benefit analysis and contingent valuation are sometimes used to convert a multidimensional set of outcomes into a single metric, usually money, in order to facilitate comparisons among risks or among policy options. Although such techniques can illuminate the choices that society must make, they cannot substitute for a deliberative process by artificially simplifying complexity. Broadly based deliberation can address whether and how to use the simplified indices that come from such techniques.

As discussed in Chapter 2, analytical procedures for summarizing information can be used in different ways so that each, although accurate, creates a different impression on audiences that have only the summary to inform them (Stern, 1991). An example is "risk ladders" that list numerous risks along a dimension, such as number of annual fatalities, which can create different impressions depending on which risks are chosen to anchor the ends of the ladder (Sandman, Weinstein, and Miller, 1994). Methods of quantitative uncertainty analysis have a similar potential for creating conflicting impressions (Finkel, 1990). Deliberative processes that involve participants with diverse perspectives on the risk decision can determine how such techniques might be appropriately used to support a risk characterization.

Deliberation can clarify the nature and extent of agreements and disagreements among participants (Gray, 1989). For instance, in situations in which uncertainties are large or data are incomplete, deliberation among scientists, public officials, and interested and affected parties can help clarify the extent to which disagreements are rooted in differences in how people or groups see the problem, how they interpret existing data, or in their values. This clarification can inform decisions about whether further analysis might help resolve the disagreements.

Deliberation can also promote mutual exchange of information and increase understanding among interested and affected parties. For example, in the South Florida case, natural and social scientists and agency personnel carefully considered issues in ecosystem management that might concern the various interested and affected parties in order to develop an analysis that might facilitate dialogue among the parties. Similarly, the Hanford Future Site Uses Working Group spent considerable time defining the common base of information that all participants wanted to consider in developing land use options. This process gave agency personnel, as well as other interested and affected parties, a richer perspective on the problems at the Hanford site and the nature of the parties' concerns. Analytic techniques such as multiattribute utility analysis (e.g., Keeney and Raiffa, 1976, von Winterfeldt and Edwards, 1986) may also be used to increase mutual understanding by clarifying which values the various participants consider important.

Deliberation also has the potential to yield more widely accepted choices, both about risk characterization and about policy. For example, dispute about an exposure assessment is less likely when the assumptions built into it have been agreed to in advance by the interested and affected parties. Similarly, deliberation may arrive at more acceptable ways to provide information. During the aftermath of the 1979 nuclear power accident at Three Mile Island, neighboring residents were highly critical and suspicious of General Public Utilities, the company that owned the plant, and of the Nuclear Regulatory Commission (NRC). When the operator proposed venting the krypton gas that had accumulated in the containment dome as a first step in gaining access to the reactor vessel, public opposition was strong. Reassurances that the risk was "minor" were not convincing. Different actors used forms of deliberation in their efforts to address the issue. Pennsylvania Governor Thornburgh conferred with two environmental organizations that represented opposition viewpoints. When they were satisfied that they understood the risks and that the risks were acceptable, the governor approved venting the krypton. The NRC held public hearings. A local mayor proposed providing nearby neighborhoods with radiation monitoring equipment and involving a team of local citizens in the design and operation of a radiation monitoring plan that it was hoped would help residents understand the risks (Gray, 1989).

Broadly based deliberation can also increase acceptance of the substantive decisions that follow risk characterization. For example, numerous case studies support the claim that disputes about siting municipal and hazardous waste disposal facilities are lessened when the interested and affected parties are made part of the decision-making team (e.g., Kraft, 1988; Heiman, 1990; Vari, Mumpower, and Reagan-Ciricione, 1993). As already noted, such participatory deliberation builds acceptability of decisions in part by fulfilling democratic norms. People are more willing to accept the results of processes they perceive as fair, balanced, and reasonable and that allow them an adequate opportunity to have a fair say. Thus, mutual agreement on the selection of technical consultants is more likely to lead to acceptable analyses (Ozawa, 1991) and, in some cases, has also reduced the number of "dueling experts" (Susskind and Cruikshank, 1987).

The successes of negotiation for regulatory rule making are largely due to appropriately broad deliberation. In this procedure, the convening agency assembles a negotiating committee comprising representatives from governmental agencies and from interested and affected parties that might otherwise challenge the rule in court. By involving these parties in the tasks leading up to risk characterization as well as in substantive negotiations, misunderstandings and disagreements about scientific knowledge are ironed out early on. Analysis tends to focus on the issues that divide the parties, and the negotiated rule better fits the understandings and matches the needs of the parties. Regulatory negotiation has not been used frequently in the U.S. government, but it seems to be gaining in popularity. Some recent evaluations of the technique have been published by Rushefsky (1991), Fiorino (1995), and Hadden (1995).

Appropriately broad deliberation may also help improve the trustworthiness of risk decision-making bodies. Risk decision makers are sometimes criticized for failure to exercise reasonable judgment; this criticism might be lessened through deliberative processes that seem equitable to the interested and affected parties. Deliberation can also help strengthen a decision maker's reputation for trustworthiness by exposing decision assumptions to testing and verification by outside parties (e.g., Jasanoff, 1990).

LIMITATIONS AND CHALLENGES

A process that combines appropriately broad deliberation with analysis makes for better informed decision participants. It can also address many procedural concerns of the interested and affected parties. However, there are limits to what better risk characterization can accomplish and major challenges associated with incorporating broad-based deliberation into the process. It is important to recognize the limits, address the challenges, and set realistic expectations.

Limitations

The best designed analytic-deliberative processes cannot eliminate all the problems and controversy associated with contentious risk decisions. They cannot guarantee acceptance of an agency's risk decision or even a risk characterization (e.g., Rosener, 1978). Controversies often reflect basic differences in values or interests: the more that is at stake and the more that values and interests conflict, the less likely it is a decision will be widely accepted. Processes that better inform the decision should not be expected to reduce these basic conflicts. Deliberation may increase understanding without narrowing the differences among parties. It may also fail to reduce conflict because some parties refuse to join the deliberative process in order to preserve the strength and legitimacy of their opposition. Or they may join the process to press strategically for delay or to shift the debate to issues that they see as more fundamental or as more advantageous to their positions. Deliberation may even exacerbate conflict or harden established positions.

Good deliberation also cannot guarantee that improved understanding will influence the final decision. Decision makers may, for a variety of reasons, ignore or reject information from an analytic-deliberative process, its conclusions about the risk, or any recommendations that may result. For example, the California Comparative Risk Project (see Appendix A), after a lengthy, expensive, and broadly participatory process, proposed an approach to setting regulatory priorities that was almost immediately rejected by the governor. Some trade organizations were dissatisfied with the project's decision to emphasize social welfare outcomes, including outcomes such as anxiety, that can only be assessed subjectively, and their view of the project's approach as downplaying "the traditional role of science" in favor of "values, opinions, fears, and anxieties" was taken up by prominent media sources (e.g., Clifford, 1994:A1). The governor and the state environmental protection agency quickly divorced themselves from the project's recommendations and rankings. Another example comes from the regulatory negotiation over disinfectant byproducts in water (Appendix A), in which the one significant interested party that elected not to participate in the negotiation appealed successfully to the U.S. Senate to pass amendments that overrode the negotiation in ways that suited its interests.

Those who organize and participate in analytic-deliberative processes need to be aware of their limitations from the outset. Participating in the process of improving understanding does not by itself guarantee that one's interests and perspectives will be taken into account in the final decision.

Good deliberation cannot redress the situation in which legal guidelines mandate that decisions be based on a different set of considerations from those that participants in an analytic-deliberative process believe appropriate. This is the case, for instance, when laws either prohibit cost considerations from being taken into account in risk decisions or require that decisions be supported by cost-benefit analyses that exclude other considerations that some of the parties believe are important. Those responsible for a risk characterization should think carefully before convening a broadly based deliberation when such constraints are likely to become an issue for participants; similarly, interested and affected parties should consider whether they want to participate in a process if their serious concerns cannot be given consideration in the decision.

Challenges

Deliberation also presents several challenges that may be addressed with careful attention to the process. One major challenge is to involve the full range of participants whose knowledge, insights, perspectives, and skills are needed for the particular task. Scientists, public officials, and interested and affected parties may each have unique expertise and a valuable perspective on an issue, so each task leading to a risk characterization may benefit from each group's participation. It may be difficult, however, to get all the important participants involved when they are needed. Sometimes, the best informed scientists are too busy to make an adequate commitment. Sometimes, some of the interested and affected parties do not have the resources—money, time of the appropriate people, or expertise—to be effective participants. For some interested and affected parties, particularly if they have few economic and social resources, achieving meaningful participation is a particularly difficult problem, even if their presence in the analytic-deliberative process is assured. These groups may be unorganized, inexperienced in regulatory policy, or unfamiliar with and inexpert in risk-related science. They may not know what kinds of information are most important to bring into the analytic-deliberative process, nor have the resources to organize that information in a form that other participants, more accustomed to science policy discussions, will recognize as authoritative. Sometimes they may have been alienated by past interactions with the agency or turned away by mountains of indigestible "information."

When such parties are particularly at risk and when they may have critical information about the risk situation, it is worthwhile for responsible organizations to arrange for technical assistance to be provided to them from sources that they trust and that can help them present their information and perspectives effectively during the process. Because there is limited experience to draw on for offering such assistance, responsible organizations should experiment and make systematic efforts to learn from the results.

It may also be difficult to know whom to involve. Trying to involve everyone who might contribute can make a process unwieldy, but methods of selecting representatives can leave the process vulnerable to the charge that it was exclusionary. This choice between openness and selection presents a dilemma with various possible solutions, depending on the situation. We discuss some methods for selecting participants later in this chapter.

Managing resources presents a different set of challenges. Questions will inevitably arise about whether the additional time and money required for considering and sometimes implementing a broader analytic-deliberative process are justified. The answers, of course, are situation specific. The amount of additional effort—if any—that is appropriate can vary greatly, as can the likely benefits. The organization responsible for characterizing a risk must decide, first when it diagnoses the situation and then in a more definitive way as the process goes on, how much analysis and what length, breadth, and type of deliberation are appropriate for the decision at hand. We discuss these issues further in Chapter 6.

Organizations need to give careful thought to the design of the decision process in order to ensure that each activity is tied in with the making of the decision. A jumble of public meetings, advisory committees, workshops, planning groups, hearings, and panels scattered throughout the process is unlikely to contribute to the risk decision and is likely to convey the impression that the organization is not interested in meaningful participation. Early and thoughtful attention to process design issues may sometimes reduce expense by showing that certain deliberative activities would be unproductive.

Judgments about resource needs are difficult and require balancing at least three kinds of consideration. First and most obviously, more analysis and deliberation have immediate costs. They use money and personnel and they risk delay, including the possibility that a responsible agency will fail to meet legal deadlines. These costs are real, and they can be substantial.

The second consideration is that truncating the analytic-deliberative process can waste more resources than it saves. An organization can become too impatient for results to allow the deliberation to proceed at its own pace, and organizational pressures may force the officials responsible for deliberative processes to end them prematurely. These pressures create the potential for illusory deliberation, a situation in which an organization formally allows for broad input but does not provide for an interchange of ideas or really consider the substance of the input. There is also a potential for false consensus—a rushed agreement on a recommendation before a deliberative group has agreed on the problem. Such failures of process can lead to bad decisions, invalidate deliberative procedures, and create resentment and mistrust as well. They can also provoke dissatisfied parties to actions, such as legal challenges, that delay decisions and increase their ultimate cost.

The third consideration is that parties that stand to benefit from delay may call for more analysis or deliberation as a strategy for achieving their ends, perhaps citing as justification the need for deliberation at every step. Not even regulatory agencies are exempt from the temptation to delay strategically. They may have various reasons for wanting to avoid making final decisions (see Graham, 1985; Dwyer, 1990). We discuss the problem of reaching closure in more detail in Chapter 5.

We are convinced that in the past, some high-profile risk decisions have suffered because not enough attention was given to the analytic-deliberative process that supported risk characterization. We believe that there will continue to be important risk decisions in which the overall process will benefit from expanded analysis and broader deliberation, even though that process imposes initial costs of time and money. It is difficult to judge correctly in advance how much analysis and deliberation are warranted in a particular new situation, especially under resource constraints. We offer some guidance on this matter in Chapter 6.

STANDARDS AND GOALS FOR DELIBERATION

Although formal techniques of risk analysis have evolved tremendously over the past 20 years, research on deliberative methods has received far less attention. Government agencies that have experience with deliberative processes usually do little to document and evaluate them, in spite of the expressed need for evaluation results and for advice based on experience (Fisher, Pavlova, and Covello, 1991; Chess, Salomone, and Hance, 1995; Chess et al., 1995; Lynn and Busenberg, 1995). Thus, much less is known about how to select deliberative methods and use them effectively in particular situations than about how to select and use analytical methods. Nonetheless, there is sufficient knowledge about the failures and successes of deliberative methods to identify some basic standards and goals. There is also enough knowledge to warrant greater explicit use of deliberative methods, as well as better documentation of their effects.

Involve the Interested and Affected Parties

Many government agencies operate with the default assumption that risk analysis and characterization are a matter for experts within the agencies and public policy officials and that special justification is needed to involve the interested and affected parties. Behavior based on this assumption may lead some of the interested and affected parties to feel disenfranchised from the regulatory process and either withdraw from the policy arena or seek unconventional ways to interfere with the process. By the time such a group attracts the agency's attention, the decision making may be too far advanced, or the agency too committed to a certain problem formulation, or trust and mutual respect so eroded as to preclude meaningful participation and deliberation. This dynamic can also be very costly in time and resources expended on protracted controversies.

We propose that government agencies operate on the opposite default assumption: that participation across the spectrum of interested and affected parties is warranted at each significant step of the analytic-deliberative process that leads to risk characterization. Particularly for government regulatory agencies that have limited public trust, it is usually wiser to err on the side of too broad rather than too narrow participation. Agencies that characterize risks should carefully and seriously assess the need for involvement of the spectrum of interested and affected parties in each step. Although the need for involvement of the parties depends on the risk situation and the particular step of the process, agencies should begin with a presumption in favor of involvement. Agencies are likely to gain facility over time in making such assessments, as well as a refined awareness of the needs and concerns of the groups that their decisions affect. Deliberations themselves may become easier as agency staff and other participants come to understand each other better.

Who Are the Interested and Affected Parties?

The term *interested and affected parties* is intentionally broad and is not meant to be limited to a legal definition of parties. The number and types of interested and affected parties will depend on the particular context of a risk situation. They may include people from diverse geographic areas, ethnic, or economic groups and organizations such as firms and local governments. The parties may include interest groups, such as trade associations, labor unions, environmental and consumer groups, and religious groups. The parties' concerns may focus on various possible forms of harm, not only mortality and morbidity, but also physical, social, economic, ecological, and moral effects.

Parties sometimes do not know that they are interested or may be affected by a risk decision unless they are informed. They may not be aware of the hazard, or their exposure to it, or the proposed mitigative actions, or the effects these actions may have on them or on things they value. Even if they are informed, they may not have the means, the power, or the level of trust necessary to participate in the decision process. In such cases, agencies have a duty not only to inform but also to facilitate the involvement of these parties, with funds where necessary.

The interested and affected parties can often be identified by considering the answers to the following questions (Chess and Hance, 1994):

Who has information and expertise that might be helpful?
Who has been involved in similar risk situations before?
Who has wanted to be involved in similar decisions before?
Who may be affected by the risk characterization?
Who may be affected but not know they are affected?
Who may be reasonably angered if they are not included?

Generally, parties that are interested or affected by a risk, or by a possible decision about risk, are candidates for participation at all the steps leading to a risk characterization.

Is Direct Participation Needed?

A key question is whether interested and affected parties should be represented by members of their own groups, or whether it is adequate (or even preferable) for them to be represented by surrogates, such as attorneys or scientific advisers. The answer is that the kinds of participation necessary depend on the particular decision and the task. The South Florida case suggests that for properly formulating a problem and arriving at a good plan for generating decision-relevant analysis, it may sometimes be sufficient to have indirect representation that makes known the affected parties' points of view and concerns, but does not include their physical presence. Direct involvement of affected parties may be essential, however, when they have local knowledge that cannot otherwise be brought into the process (Mazmanian and Nienaber, 1979). For example, farm-workers should almost certainly be consulted in planning the analysis for a decision on regulating a pesticide, because they have the best knowledge of the actual conditions of application that determine their exposure (Vaughan, 1993b).

To increase the legitimacy of decisions, broad, direct participation is important, though there may be limited conditions in which indirect participation is adequate. Many European governments achieve a high level of acceptance of environmental policies by consulting designated representatives of major communities of interest (e.g., corporate, labor, and perhaps environmentalist), without formal mechanisms of open public comment or participation (e.g., Jasanoff, 1986). Because of the more open and skeptical political culture in the United States and the history of skepticism about many government agencies' decisions, this sort of indirect participation procedure is likely to be of less value in this country—but there can be exceptions. For example, the South Florida ecosystem restoration project used a strategy of indirect representation—using social scientists as surrogates for affected groups—that appears to have been successful in getting a range of those groups' concerns addressed in the analysis.

Some U.S. regulatory organizations, such as the National Transportation Safety Board (NTSB), analyze and characterize risks (e.g., of aircraft accidents) in ways that appear to have wide acceptance despite the fact that the analyses are made mainly by experts drawn from affected organizations, without formal input from "the public." Thus, it may be possible to achieve legitimacy without direct participation, either by doing what is widely perceived to be a good job of risk minimization (as the NTSB has so far been able to do) or by involving surrogates thoroughly enough to satisfy the interested parties of the fairness of the process. Both of these strategies, however, can quickly come undone with one or two controversial or embarrassing decisions. Should that happen, agencies may need to reestablish legitimacy by expanding openness to and consideration of outside concerns.

A difficult issue in participation is how to consider the interests and perspectives of parties that cannot be involved by direct participation. These parties may include children, the disabled, and future generations, as well as nonhuman species and ecosystems. Since it may be necessary to represent these perspectives to achieve a full understanding of risk and an acceptable risk characterization, some form of indirect representation is the only option.

Selecting Participants

Four key considerations should be kept in mind in selecting participants: that the participation is sufficiently broad; that the selection process is fair and perceived as fair; that participants who presumably represent interested and affected parties are acceptable to those parties as representatives; and that the participants bring to the process the kinds of knowledge, experience, and perspectives that are needed for the deliberation at hand. It is often wise for a regulatory agency to enlist outside help in choosing participants.

There are several selection strategies. One is self-selection. This is the standard procedure in public hearings and notice-and-comment rule-making: an organization or government agency makes a public announcement, and anyone who wishes may participate. Although this approach is fair in the sense of allowing equal opportunity, it has some well-known limitations. It favors interests that are already organized and those that have enough resources to monitor announcements, mobilize interest group members, submit comments, or participate in other ways. It does not deal with the problem of participation by parties that do not yet realize they may be affected. And it may result in too few or too many participants (English et al., 1993). Very widespread and intense participation can become a problem if an organization does not have enough time or personnel to consider all ideas seriously and therefore appears to have invited participation only to ignore it.

A second strategy is to select individuals specifically to represent each interested and affected party. This approach raises several questions: Who determines who is or represents an interested or affected party? In particular, if some of the parties are not organized, who identifies them as groups and selects individuals to represent them? Should all the parties be given equal weight? As noted above, how can silent parties, such as future generations, be represented (English et al., 1993)? Clearly, the challenge of representation as a strategy is to identify all the relevant parties and represent them in ways that they consider adequate.

A third selection strategy relies on sampling techniques. This approach is sometimes used to ensure that participants represent the demographic makeup of a geographic area. There are sophisticated techniques that can stratify a population according to appropriate criteria (e.g., sensitivity or vulnerability, location, age, race) and select a group that is representative on the criteria considered important for the decision. This approach also has difficulties, however. First, random selection of a small group may not fairly reflect the range of interests in a risk situation. Second, random sampling presumes that everyone should have equal say, but there are sometimes strong moral arguments for paying special attention to particular groups, such as children. Third, the individuals selected may not be equal in their interest, in their time to participate, or in their ability to understand the issues, so that some of the parties may get better representation than others. Some managers try to surmount this problem by paying randomly selected citizens for their time and making special efforts to educate all the participants on the issue (e.g., Renn et al., 1993).

There have been various efforts to balance the strengths and limitations of the above approaches. For example, in a transportation planning process convened in Boulder, Colorado, 147 randomly selected residents gave input to a smaller advisory group through a series of telephone interviews, in-person interviews, and mailed surveys, thus broadening the range of perspectives included in the deliberative process (Kathlene and Martin, 1991). In developing water quality regulations for New Jersey, a self-selected task force representing interested and affected parties negotiated a reduction in the size of the group while working out a balance of representation of various interests; meanwhile, the responsible agency invited other interested citizens to informal meetings (Chess, 1989).

The three-step method being used in Europe (Renn et al., 1993), uses different selection processes for different stages of the process. First, representatives of known interested and affected parties raise issues. Second, scientific experts provide judgments on those issues (in the form of risk estimates). Third, randomly selected citizens' panels deliberate on the decision, using the information from the first two stages of the process. Regulatory negotiations in the United States have used different tiers of participants, with a small group of key participants conducting the negotiations, a larger group in the second tier observing and conferring with the negotiators, and a third tier of participants reviewing the agreement before it is ratified (Gray, 1989).

Timing Participation

Participation is needed "early" in decision processes, but we prefer to call for the consideration of broad participation throughout the process, or in all its significant steps. Experience suggests that including a full range of perspectives is especially important in problem formulation, because problem formulation profoundly shapes how risks are understood. It is also important when options are being considered, for the same reason. But the full range of perspectives is also needed for other purposes. This argument may seem least obvious for the knowledge-gathering step, which is typically envisioned as the preserve of scientific and technical experts. External input can be important in this step as a source of expertise on local conditions and social and organizational factors that affect exposures to hazards. Such expertise is needed to keep risk assessments and risk characterizations from adopting implausible or false assumptions.

Listening to the Participants

Many critics have described public participation as consisting only of token efforts at soliciting input (e.g., Arnstein, 1969; Chess and Salomone, 1992). A review of research on citizen advisory committees suggests, however, that the intent of the convening agency can strongly influence outcomes, with some agencies using the process mainly to fulfill legal obligations or to persuade outsiders to accept its decision, while others are able to use it to get valuable input to policy (Lynn and Busenburg, 1995).

Explicitly Address External Constraints

Government agency officials may see proposals to involve outsiders in their deliberations as an invitation to abdicate their legal responsibility to act. Experience shows, however, that agencies can creatively use external input to improve the analytic-deliberative process without running afoul of their legislative mandates. Deliberation can be used, for instance, to inform discretionary decisions about how to gather knowledge, decisions that are unconstrained by law.

Government agencies that want to broaden participation in their deliberations should carefully examine the legal limits of their discretion and design a process that is defensible within those limits. Statutes that mandate a particular form of public participation may be seen as precluding other forms of deliberation, but often they do not, although there are important limitations. At present, for example, the Federal Facilities Advisory Committee Act limits direct interactions between agencies and outside interested and affected parties. While developed to prevent collusion, this restriction also inhibits innovative deliberative processes (Crowfoot and Wollendeck, 1990). Mandated regulatory timelines can also be a serious constraint on deliberation and may lead agency officials to limit the length of a deliberative process. Time schedules can present a serious problem for lower level officials in agencies who are asked to organize deliberative processes because nontraditional approaches to deliberation are perceived as time-consuming (Crowfoot and Wollendeck, 1990).

Conveners of deliberative processes should clearly and explicitly inform participants at the outset about any constraints on the process and on how the agency can, or is likely to, use their input. Is it willing and able to commit necessary resources? Will it be represented in the deliberatory process by personnel with sufficient authority to make commitments? How much impact will the deliberative process have on the risk characterization? Are there aspects of the risk characterization in which the parties' influence will be restricted? Are there legal restrictions on what can be considered in making the decision? Addressing such questions can avert damaging misunderstandings. Agency officials should make it clear to the participants what the process is intended to accomplish and what their roles and responsibilities are. Ideally, the participants should be involved in designing the process.

Legislative mandates, procedural restrictions, and agency culture are not the only important external constraints on deliberation. Interested and affected parties may be a source of constraints, too (Bingham, 1986). For example, as noted above, some parties may perceive an interest in delaying risk characterization or in prolonging the analytic process in order to delay a decision. They may therefore welcome an opportunity to deliberate and use it to create delays. Involvement in legal actions may constrain the participation of some parties, or they may see it as doing so. And conflicting agendas between parties to a deliberative process may make deliberation difficult. Agencies should consider these possibilities carefully when planning deliberations.

Strive for Fairness in the Process

Many disputes around risk have been over the fairness of the process that generated the risk estimates on which a decision was based. Procedural justice issues about who participates in decisions, and how, have caused as much or more conflict as the risk estimates themselves (Clarke and Short, 1993). Thus, deliberative processes should strive for fairness.

It is important to recognize that notions of fairness change over time: For example, it is now unfair, unethical, and illegal to discriminate against women in ways that were socially condoned 100 years ago. Also, judgments about the fairness of a deliberative process or its products are affected by people's past experiences, access to power and influence, and cultural backgrounds (Rayner, 1992). Fairness also depends on judgments about how well a process has respected generally accepted rights, such as to participation and to informed consent. Thus, fairness is partly in the eye of the beholder, making it the subject of intense debate. For those designing a deliberative process, the fundamental challenge is to design a process that the interested and affected parties will see as fair, as well as effective.

A major point to consider is the relative power of interested and affected parties. For example, to what extent should special efforts be made to give access and influence to parties that have been traditionally disenfranchised? The concept of a level playing field is one that many people accept conceptually, but it can generate serious conflict in its implementation. Power can be made more equal by providing access to expertise, information, opportunities for input, and other resources to parties that normally lack them; however, doing so can tax an agency's budget. Decisions about resource allocation should be guided by the need for the deliberation to be balanced—that is, to include the full range of knowledge, insights, and perspectives needed to characterize the risks—and to be procedurally acceptable to the interested and affected parties.

Plan for Flexibility and Iteration

Careful management of the deliberative process, including designing in flexibility, is critical to its success. First, deliberative processes require all the usual kinds of management: development of timelines, delegation of responsibilities, resource allocation, and coordination. Failures of deliberative processes have been traced to failure to plan (e.g., Nakamura, Church, and Cooper, 1991), and poor management may increase participants' frustration and cynicism (Freudenberg, 1983).

Second, managers should take into account that deliberations sometimes result in a call to revisit past decisions—for example, to gather new data in order to summarize knowledge better. They should expect such requests to arise and consider procedures for responding to them.

Third, managers should consider the role of deliberation in each of the steps leading to a decision, from problem formulation through knowledge generation and summarization. They should consider how various values and interests might affect each task and how to use deliberation to ensure that concerns are considered at each step in ways that are credible to the interested and affected parties.

Fourth, agency planners should address and, when appropriate, resist the temptation to standardize procedures. Government agencies typically define standard procedures for deliberation, such as the routines of notice-and-comment rulemaking or of participation in public hearings. Routines are advisable when an agency faces a series of similar decisions. But different situations may call for different processes, and even frequently repeated decisions sometimes need a change of procedure. It is not generally wise to enforce a standard procedure for deliberation across a variety of decision situations. With frequently repeated decisions, it may be appropriate to conduct a broadly based deliberation to arrive at a procedure for characterizing risks and then to implement that procedure routinely for many decisions until it is time to reconsider the procedure, in a second broadly based deliberation. We discuss such issues in more detail in Chapter 5. It may be useful for a process that begins with one deliberative method to add or shift to others over time. (e.g., Mazmanian and Nienaber, 1979) For example, an internal agency task force, a task force of interested and affected parties, informal meetings with interested and affected parties, and formal reviews of draft regulations may take place concurrently or sequentially (Chess, 1989). The most appropriate form of deliberation may even depend on geography: decisions involving interested and affected parties that are widely dispersed (e.g., a federal regulation on emissions into a river) may need to structure deliberation differently from one involving interested and affected parties that are in closer proximity (e.g., a landfill). Various other factors, including agency resources, the extent of controversy, the complexity of the problem, and the preferences of interested and affected parties, will influence the choices of deliberative methods (English et al., 1993). It is appropriate to consider such factors in adopting or designing a deliberative process.

Perhaps the key to successful deliberation is explicit attention to process design: a process that the participants agree to at the outset has the best chance of being acceptable. A prudent strategy for agencies is to adopt a willingness to cooperate with interested and affected parties in reaching agreements on the deliberative methods for specific cases.

Recognize the Roles of the Responsible Organization

The organization that organizes an analytic-deliberative process has a number of roles to play in facilitating deliberation. As the *convener*, it has initial responsibility for diagnosing the situation and making initial estimates of time and resource needs, of who should be involved both within and outside the organization, and of the tasks that need to be accomplished. We discuss diagnosis in more detail in Chapter 6. Staff also need to seek support of key decision makers for the deliberative process. The organization is also responsible for discussing the initial plans for the activity with the initial participants.

The responsible government agency or other organization is also a *coparticipant*, with a legitimate interest and, perhaps a legal mandate to be involved in the risk characterization and the goal of reaching a fair and wise policy decision. An agency that is responsible for managing a risk is not a neutral party. Given agency expertise in technical analysis and sensitivity to the legal and political constraints of decision making, its staff may feel it knows what the best decision is and may be tempted to try to influence the outcome of the deliberation. An agency's expertise and its power over the decision need to be clearly stated and acknowledged by all participants. But it is the agency's responsibility to offer sound reasons if it chooses to ignore the results of a deliberative process.

The agency usually acts as *overseer* of the deliberation, the party that usually works to break deadlocks and to reach closure. Unless it designates an oversight body for the deliberative process, the agency retains responsibility to resolve fundamental disagreements over the operation of the process. The agency should, however, make serious efforts to distinguish its role as a party to the deliberation from its role as overseer of the process.

Use Appropriate Methods

Practitioners have developed a great variety of techniques that can be used in the deliberations that contribute to informing risk decisions. These are described in an extensive literature, and there have been attempts to catalog or classify them (e.g., Creighton and Delli Priscolli, 1983; English et al., 1993). However, there is no rigorous or generally accepted classification scheme, and it is not possible to predict which deliberative method will work most effectively in any given situation. Deliberative methods are merely tools. Results will depend less on the tool and more on its users and the setting in which it is used. For example, although public meetings are notorious for facilitating posturing rather than deliberation, they have also been successful in certain situations (e.g., Rosener, 1982; Hadden, 1989; Webler and Renn, 1995). Citizens advisory committees have widely different levels of impact, depending on the intentions and expectations of the agency that organizes them (Anderson, 1986, Lynn and Busenberg, 1995; Lynn and Kartez, 1995). The history of an issue, level of conflict, scientific data, and existing power dynamics may also influence outcome as much as the method. For example, years of hostility among interested and affected parties, or between parties and an organization, can undermine deliberation, regardless of the method used.

The choice of deliberative methods requires diagnosing the risk situation and the nature of the knowledge needed, including the needs of the parties, the technical complexity and history of the issue, the extent of agency commitment, the availability of expertise in deliberative methods, and available resources. Deliberative processes also need to be sufficiently flexible to allow for mid-course corrections (English et al., 1993). And as already noted, the deliberative approach may need to be tailored to the needs of the specific situation.

Appendix B describes several important methods of deliberation and public participation. It notes some of the strengths and possible limitations of each, and ways of compensating for the latter, and it refers to sources of more detailed analyses of experience with each technique. We list these methods to suggest the variety of possibilities available to agencies, but we do not consider the list exhaustive, and we do not recommend general use of any particular method or technique. In our judgment, the choice of a deliberative technique depends on the risk situation and on the particular risk characterization task. Within the course of a single effort at risk analysis and characterization, an agency may successfully use several different deliberative techniques or develop new variants, using different techniques to accomplish different tasks.

4

Analysis

We use the term *analysis* to refer to ways of building understanding by systematically applying specific theories and methods that have been developed within communities of expertise, such as those of the natural science, social science, engineering, decision science, logic, mathematics, and law. Risk analysis, an activity that applies analytic techniques to the understanding of risks, has grown rapidly since its beginning in the 1950s. It involves estimating the likelihood of occurrence and possible severity of particular kinds of harm. Analysis can also be used to examine risk problems to characterize their history and analyze possible outcomes of different decisions, strategies or policies. Risk analysis can be qualitative as well as quantitative; in fact, for some important elements of risk, no valid method of quantification is available.

Analytic techniques are essential for understanding risk, and many useful volumes have been written about them (e.g., Raiffa, 1968; U.S. Nuclear Regulatory Commission, 1975; Lewis et al., 1975; Fischhoff et al., 1981; von Winterfeldt and Edwards, 1986; Crouch and Wilson, 1982; Travis, 1988; Cohrssen and Covello, 1989; Morgan and Henrion, 1990; Rodricks, 1992; Royal Society Study Group, 1992; Suter, 1993; National Research Council, 1994a). For this reason, our treatment of analytic techniques is brief. Chapter 2 has pointed to the need to apply analytic techniques more broadly, so as to expand the aspects of risk that are given careful scientific attention. This chapter discusses the general principles and purposes of analysis and addresses two substantive analytical issues that have received much attention in recent discussions of risk characterization: the appropriate use of analytic techniques to reduce the multidimensionality of risk and the analysis of uncertainty.

Risk analyses usually address such basic questions as: What can go wrong? How likely is it? What are the consequences? How certain is this knowledge? (see Kaplan and Garrick, 1981). Although these questions are most often asked only about risks to human health and safety and the environment, they can in principle be asked about the full range of harms that concern interested and affected parties and public officials. We emphasize that analysis can be used for social questions about risk, including potential economic, social, political, and cultural harms; the design of messages synthesizing the results of analyses; and the design and evaluation of procedures for broadly based deliberation. Analysis therefore may involve more than the tools of the natural sciences and more than quantification.

Methods for quantitative analysis include collection and evaluation of observational or archival data, experimental studies, epidemiological and econometric analysis, survey research, and the development of predictive models of the physical or social phenomena affected by the risk. Methods for qualitative analysis include systematic clinical and field observation, logical inference from historical and comparative studies, inference from legal precedent, ethnographic interviewing, and the application of principles of ethics. Although the bulk of the effort in developing methods of risk analysis has been addressed to quantitative methods, critical aspects of risk frequently require qualitative evaluation.

PURPOSES AND CHALLENGES OF ANALYSIS

Analysis is essential to the risk decision process because it is the best source of reliable, replicable information about hazards and exposures and options for addressing them. Analysis, in quantitative form when appropriate data and methods are available, offers a window on the relative magnitude of hazards and exposures. Relevant analysis, in quantitative or qualitative form, strengthens the knowledge base for deliberations, both about how to deal with hazards and about how to better inform risk decisions. Analysis can clarify issues by identifying the likely results of decisions, the implications of options, and previously unrecognized potential dangers. It can enable all parties to reach agreement on some issues and focus further discussion on areas of disagreement. It can provide a basis for selecting among positions without regard to who favors those positions. And it illuminates the decision options that are available when choices must be made with incomplete information, under uncertainty, and with strong and opposite positions having been declared.

Analysis, like deliberation, needs to be much more extensive in some decision situations than in others. It is almost always possible to consider conducting more detailed analysis so that a risk decision can be better informed. But like additional deliberation, additional analysis requires time and other resources. Judgments about the appropriateness of conducting analysis are very much part of the analytic-deliberative process: the possibility of doing additional useful analysis does not necessarily require that it be carried out.

Without good analysis, deliberative processes can arrive at agreements that are unwise or not feasible. For example, the U.S. government negotiated an agreement in 1989 to clean up the Hanford, Washington, nuclear weapons site by 2018 because "thirty years seemed like a reasonable length of time to complete the cleanup" (Blush and Heitman, 1995:ES-2). But the agreement included milestones, including one for removing tritium from groundwater that may not be met because no technology yet exists to accomplish the task (U.S. General Accounting Office, 1995). Analysis of the proposed agreements from the standpoint of technical feasibility might have led to a more realistic commitment.

Although analysis is most commonly associated with the task of gathering and interpreting data, it also provides critical input to the other steps leading to risk characterization. It can help define problems. For example, analysis of chemical processes in the atmosphere first defined the problem of stratospheric ozone depletion and predicted that it would occur as a result of anthropogenic releases of chlorofluorocarbons (Rowland and Molina, 1974). It can generate options. One example is the so-called geoengineering approach to responding to the threat of climate change (Committee on Science, Engineering, and Public Policy, 1992). And it can help summarize information, for example, by finding accurate and effective ways of presenting uncertainty.

Analytic approaches are increasingly being used to summarize knowledge. These include techniques for clear graphic presentation of data that can be of great use for understanding the many factors relevant to a decision. However, good presentation without a correspondingly high quality of substance can mislead decision participants and subvert the role of analysis. Similarly, other new decision support systems, including integrated database management and modeling, provide opportunities for improving the ability to perform, summarize, and communicate analysis. Effective decision support systems can allow analysts to access and evaluate data, in some cases in real time (e.g., for hurricane, flood, or pollutant spill evaluations); test predictive models; evaluate management and decision options; perform uncertainty analyses; and identify data and research needs to improve predictions.

Quantitative models to organize and interpret data are particularly important to risk characterization. In some fields, such as ecological risk characterization, analyses are sometimes based largely on conceptual models. Models provide a framework that defines the relationships that are valuable to study and specify how measured quantities are to be interpreted in relation to the real world. Models simplify the world and can therefore provide clear responses to policy questions. But they also present analysts with a tradeoff between the needs for simplicity and for verisimilitude. Incorporating more real-world components and processes can lead to more realistic representations, but complex models can require analysts to make many estimates, and may exceed analysts' ability to understand how the model operates and therefore to obtain meaningful insights. Simpler models provide clearer and possibly better analysis, but may omit or misrepresent some critical processes or components; there are justifications for different approaches to making the tradeoff (see, e.g., Weaver, 1948; Simon, 1982; Beck, 1987; Jefferys and Berger, 1992; Morgan and Henrion, 1990:Chap. 11). One method seeks a flexible, hierarchial, and step-wise approach to complexity. Initial model formulations are simple, attempting to frame, scope, and bound possible risks, thereby helping to identify whether and how more sophisticated analysis should be pursued. More detailed models and analyses are then developed, allowing for comparisons across levels of complexity and conceptual representation.

Models and other decision support systems also may help meet the challenge of integrating analysis with deliberation by enabling a wide range of interested parties to participate in a more sophisticated and better informed way in the analytic-deliberative process. When the underlying model and data inputs have been developed in a scientifically sound and an open and inclusive manner that inspires trust and support among participants, they can serve as a basis and focal point for joint investigation and evaluation of alternatives by all the parties to a decision. If the data and models are not understandable by participants, there is a potential for specialists to use them to manipulate the understanding of nonexperts, and for them to be perceived as manipulative.

STANDARDS FOR GOOD ANALYSIS

Good quantitative analysis has several characteristic features:

- It is consistent with state-of-the-art scientific knowledge.
- Any assumptions used are clearly explained, used consistently, and tested for reasonableness.
- The analysis is checked for accuracy (e.g., of calculations).
- Unnecessary assumptions are removed before the final analysis is

reported, after checking to ensure that the removed assumptions do not affect the results.

• Any models used for calculation are well defined and, ideally, validated by testing against experimental results and observational data.

• Data sources are identified in such a way that the data can be obtained by anyone interested in checking them.

• Calculations are presented in such a form that they can be checked by others interested in verifying the results.

• Uncertainties are indicated, including those in data, models, parameters, and calculations.

• Results are discussed clearly, indicating what conclusions they can support.

Although all these standards are reasonable, often they are not met in practice. Analysts may uncritically select assumptions that are unreasonable. They may choose, but not explain, key assumptions that substantially determine the outcome. They may even be unaware of assumptions that are implicit in the models they use. They may adopt models that are easy to use but have inherent weaknesses. They may neglect model validation because of time pressures. They may use data without checking the source and quality. They may not mention uncertainties because they are difficult to estimate, undermine the certitude with which the results can be presented, or even invalidate the analysis. They may neglect balance in an effort to strengthen their conclusions.

Good qualitative analysis has many of the same features as good quantitative analysis, but it faces greater burdens. Because it tends to have less well-established procedures, qualitative analysis tends to be more difficult to validate, more subject to opinion, and more easily discredited by skeptics. However, some of the issues most important to interested and affected parties—such as issues of informed consent and some equity issues—are only treatable by qualitative analysis. It is a challenge for researchers as well as analysts to develop reasonable standards for qualitative analysis.

For both quantitative and qualitative risk analysis, technical adequacy is a necessary but not sufficient characteristic: analysis must also be relevant to the given risk decision. First, the questions to be addressed must be appropriate for the available analytic techniques and must be ones for which information exists. An analyst often can be most helpful by identifying questions that cannot be answered with available information unless reframed. Second, the analysis should detail the limits of current knowledge, identify which factors have been included and excluded, and summarize the uncertainties associated with its results. Third, analysis should respond to the needs and expectations of the interested and affected parties. Fourth, analysis should address the issues that need to be resolved for the decision. Finally, analysis should be independently reviewed as to its assumptions, calculations, logic, results, and interpretations. This point is particularly important and often neglected. A review of what conclusions can be drawn is critical, since it is the conclusions that form the basis of a risk decision.

ANALYSIS TO REDUCE THE COMPLEXITY OF RISK

A great variety of analytic techniques exist for reducing the complexity of risk. We do not comment on specific ones, but focus instead on how such techniques can be appropriately integrated into the process that results in a risk characterization. We focus especially on the class of techniques, including those of benefit-cost analysis and multiattribute utility analysis, that aims to reduce risk to a single dimension as an aid to priority setting and decision making.

Chapter 2 emphasizes the multidimensional nature of risk and its importance for understanding and coping with risks. This complexity raises several difficult questions for risk analysis, among them the following:

• Which of the many dimensions of a particular risk are relevant to the decision at hand? For which should efforts be made to conduct quantitative analysis? For which should analysis be qualitative? Which dimensions do not need to be analyzed?

• Are there reliable and valid techniques for estimating the various nonhealth outcomes of concern, such as ecological effects, social effects, and effects on future generations?

• Which dimensions of a risk are important, and to whom? How important? How does one know?

• Is it appropriate to aggregate different dimensions of risk into a single overall measure of the magnitude of the risk? Are there reliable and valid methods that can be used for such aggregation?

• If there are no adequate methods for aggregating the dimensions of the risk, what methods should be used to set priorities for action among different hazards and risks?

Risk analysts are aware of these issues and have attempted to develop analytical techniques to address them. There are specialized techniques for analyzing particular dimensions of risk, such as ecological risks (e.g., Harwell et al., 1990; Bartell, Garner, and O'Neill, 1992; Kopp and Smith, 1993; Suter, 1993), certain social and economic effects (e.g., Finsterbusch and Wolf, 1981; Finsterbusch, Llewellyn, and Wolf, 1984; Greenberg and

Hughes, 1993), distributional equity (e.g., Zeckhauser, 1975; Anderson, 1988; Leigh, 1989; Ellis, 1993), and intergenerational equity (e.g., Viscusi and Moore, 1989; Cropper, Aydede, and Portney, 1994). There are also techniques for addressing several dimensions of risk at once to try to simplify the understanding of risk—by combining many dimensions into one. Some of these techniques convert deaths, illnesses, and nonhealth outcomes into monetary units for use in cost-benefit analysis (for a review covering several methods used in economics, see Cropper and Oates, 1992). Some aim to arrive at a nonmonetary, single-dimensional summary, expressed, for example, as an overall indicator of health risk or quality of life, as a basis for making comparisons and setting priorities (e.g., Olsen, Melber, and Merwin, 1981). Others, such as the techniques of multiattribute utility analysis, allow for different ways of reducing the dimensionality of risk depending on value priorities specified by the users (e.g., Keeney and Raiffa, 1976; Edwards and Newman, 1982; see Appendix A for one example of an application). And there are techniques for making quantitative comparisons between risks that vary in their uncertainty profiles (e.g., Finkel, 1990).

Such analytic techniques have been developed to illuminate and try to bring rationality to difficult choices between alternatives whose risks (and benefits) differ qualitatively as well as quantitatively. They respond to the need of decision makers for better ways to take the various dimensions of a choice into account and for a rational and defensible basis for making decisions. Government agencies may also use the techniques to routinize their decision processes and to meet legal tests regarding arbitrariness and capriciousness.

There are two chief strengths of such analytical techniques: they require analysts to pay careful attention to several dimensions of risk and, in the course of deciding on how to aggregate across dimensions, the techniques may elicit careful deliberation about the relationships and tradeoffs among the dimensions. Because of these strengths, such techniques can be valuable aids in understanding risk. They can make tradeoffs clearer and show what decisions would follow from accepting particular value choices.

The techniques also have significant dangers and pitfalls associated with their goal of simplifying an inherently multidimensional problem and with their use not only to inform, but also to help make decisions. Techniques that aim to simplify risk necessarily embed value choices, some of them highly contentious. Among others, they embed a choice to set risks to all individuals equal or to treat some kinds of people, such as children or people who are involuntarily exposed, as more worth protecting than others. They embed choices about whether to discount future risks and, if so, by how much. They embed choices about how to weigh risks to natural habitats against risks to economic activity, risks to human health against principles of informed consent, and so forth. In addition, they involve making a judgment that all the dimensions of risk that are relevant to the decision at hand have been considered. The values associated with each of these judgments are built into the analysis, but some of the judgments made in any given instance may not be widely accepted in the society. Thus, there are likely to be people who do not accept the judgments and value choices embedded in any particular analysis.

Because of these dangers and pitfalls, we express caution about the use of analytic techniques to simplify risk. These techniques can be helpful, but they should be handled with care and should not be used to dominate decision making. Similar concerns have been expressed by many others (e.g., Lave, 1981; Dietz, 1987, 1994; National Research Council, 1989; Jasanoff, 1993; Fischhoff, 1994, 1995). Recently, a broad group of economists reviewed the use of benefit-cost analysis in environmental, health, and safety regulation and reached similar conclusions (Arrow et al., 1996:3,7,10):

> Benefit-cost analysis is neither necessary nor sufficient for designing sensible public policy. If properly done, it can be very helpful to agencies in the decisionmaking process. . . . There may be factors other than benefits and costs that agencies will want to weigh in decisions, such as equity within and across generations. . . . Care should be taken to assure that quantitative factors do not dominate important qualitative factors in decisionmaking.

Our caution derives not from the fact that these techniques require their practitioners to exercise judgment—judgment is involved in all techniques that simplify complex realities in the service of decision making. The danger lies in using judgments that are implicit in analytic techniques but are made without broad-based deliberation, as substitutes for that deliberation. It lies in acting as if values are not embedded in the analyses or as if some particular analytic technique can be assumed in advance to yield the best or most trustworthy understanding of a risk situation. Government agencies may be strongly tempted to use analytic techniques as substitutes for informed and appropriately broad-based deliberation in weighing conflicting values because of their need for routine and legally defensible decision procedures. They should resist this temptation.

Analytic techniques for simplifying risk may aid the analytic-deliberative process or interfere with it. Research does not offer a basis for definitive guidance as to how to make these techniques helpful. Our experience and our reading of the case material suggest that the key is that the deliberative process should help shape the analysis, determining which particular techniques are used and how their results are interpreted. Especially when the decision at hand is highly controversial and when strong values and interests may come into conflict, it is important that the spectrum of scientists, public officials, and interested and affected parties come to agreement in advance on which techniques of simplification, if any, will be used and what they will be used for, and that they have the opportunity to examine the way the techniques are being used, to question the analysts, and to demand that the analysis be varied in ways that they believe will illuminate their deliberations. In short, there should be appropriately broad-based deliberation and iteration concerning the use of these techniques, just as with other risk analytic techniques. Without such feedbacks, it is more likely that the interests that appear to lose on the basis of the analysis will criticize the analytic technique as biased, thus defeating the hope that analysis will yield rational, defensible, and legitimate decisions.

Some people may object that nonexperts are incapable of making competent decisions about complex analytical techniques that they do not understand. But the fact that they may not understand the techniques is precisely the reason that the analysis must be responsive to the information needs of the interested and affected parties, as determined by the deliberative process. So long as decision participants understand which value assumptions underlie an analysis, the analysis can serve the decision. To the extent that the value assumptions become opaque, as can occur when analysis uses unnecessarily sophisticated mathematical techniques or when value assumptions are hidden in the details of a model, the analysis begins to take over the decision. Participants who do not know how value choices are affecting the analytic outputs are likely to become suspicious, especially if there is a history of distrust among the parties. Such a situation may cause more difficulties than it avoids.

We conclude that analytic techniques for simplifying risk should be treated like other analytic techniques used to inform risk decisions. That is, decisions about using them, refining them, and interpreting their results should be made as part of an appropriately broad-based analytic-deliberative process involving not only analytic experts, but also the public officials and interested and affected parties whose decisions the techniques are intended to inform.

These conclusions have implications for a collection of recent legislative proposals and agency guidances that call for using analytic techniques of benefit-cost analysis or risk analysis as the sole or primary basis for making "comparative risk" judgments or for "risk-based decision making" (a recent prominent example is in U.S. Environmental Protection Agency, 1993j). These proposals rely on analytic techniques that reduce risk to a single dimension, such as dollars or statistically expected cancer cases, as a way to make public policy decisions. They rest on two presumptions: that an available analytic technique can make such a reduction in a way that is scientifically defensible and can achieve wide social acceptance and that decisions made by using a one-dimensional scaling of risk will be socially acceptable. Like much else in risk characterization, the appropriateness of these presumptions is situation-specific. There may be situations in which the presumptions are appropriate, but they are not so in the general case. In particular, for the reasons given above, we do not believe they are appropriate for many of the highly controversial choices for which these proposals are being promoted.

We understand the need for rational, defensible procedures for making risk decisions, but we warn against adopting standard procedures that make the values and interests at stake less transparent to decision participants. Adopting such procedures may simply shift the ground of controversy from the values at stake to the arcane details of benefit-cost analysis or some other complex analytic technique. Such a shift would not, in our judgment, improve understanding of risk. At worst, it might further erode trust in already suspect government agencies.

We believe that techniques for simplifying risk may have great value for improving risk characterization and decision making if they are used carefully, in the context of an analytic-deliberative process. We warn strongly, however, against adopting them as a routine basis for decision making in the absence of evidence that they can improve present procedures. It would be worthwhile to experiment with the use these techniques in particular areas of risk decision making where they seem likely to be helpful and to carefully evaluate the effects of their use on understanding and on the decision-making process. It would also be worthwhile to experiment further with deliberative techniques for priority setting, in which an appropriately broad-based process considers information from analyses of the various dimensions of a risk and information from the application of analytic techniques that seek to simplify risk.

THE ANALYSIS OF UNCERTAINTY

Much attention has been recently given to quantitative, analytic procedures for describing uncertainty in risk characterizations (e.g., Finkel, 1990; Morgan and Henrion, 1990; National Research Council, 1994a; Browner, 1995). We discuss this topic in some detail because it illustrates the strengths and limitations of analysis and the need to combine it with deliberation.

The uncertainty of risk estimates and the interpretation of uncertainty have become a frequent focus of controversy. Uncertainty commonly surrounds the likelihood, magnitude, distribution, and implications of risks. Uncertainties may be due to random variations and chance outcomes in the physical world, sometimes referred to as *aleatory uncertainty*, and to lack of knowledge about the world, referred to as *epistemic uncertainty*. Sometimes, scientists may not know which of two models of a risk-generating process is applicable. Such situations are sometimes referred to as presenting *indeterminacy*. When uncertainty is present but unrecognized, it is simply referred to as *ignorance*. This last case is the most worrisome, as it can result in mischaracterization of risk that systematically underestimates uncertainty, with potentially serious implications for the decision process.

When uncertainty is recognizable and quantifiable, the language of probability can be used to describe it. Objective or frequency-based probability measures can describe aleatory uncertainties associated with randomness, and subjective probability measures (based on expert opinion) can describe epistemic uncertainties associated with the lack of knowledge. Sometimes, however, uncertainty is recognized but cannot be measured, quantified, or expressed in statistical terms. For instance, the economic impact of global climate change may be greatly affected by the future forms and structures of economic organization in different parts of the world, yet uncertainty about them 100, 50, or even 20 years from now is great, and extremely challenging to quantify. Similar arguments hold for many assessments of risks far into the future, such as those for radioactive waste repositories where risks are computed over design periods of 1,000 or 10,000 years. The uncertainty, especially regarding human intrusion into a repository over a 10,000-year time span, is such that "it is not possible to make scientifically supportable predictions of the probability" of such an intrusion (National Research Council, 1995:11).

Three hypothetical descriptions of risk can illustrate the prevalence and importance of the different types of uncertainty in risk characterization. Consider these three risks: a 1-in-100 chance of a river overflowing its levee in a given year with a given impact on life and property; a 1-in-10,000 chance of a volcano erupting near the proposed waste repository at Yucca Mountain in the next 10,000 years, resulting in the release of a given quantity of radioactive material; and a 1-in-1,000,000 chance of an individual contracting a fatal cancer over his or her lifetime due to a chemical exposure. Even if each of these probabilities of occurrence and impact were known with certainty, the precise realizations of the risks (e.g., when, where, to whom, and how severe the actual harm) would still be random and thus inherently uncertain. An understanding of this inherent, aleatory uncertainty is fundamental to risk characterization.

Furthermore, in each of these (and most other) cases, the probabilities of occurrence and impact are not known with certainty; they are usually highly uncertain. In the case of the river levee, the probability of occurrence may have been estimated on the basis of recent or historical streamflow records, but those records may be of limited duration or completeness and thus may not accurately represent the longer historical record. This possibility creates epistemic uncertainty. In addition, the underlying statistical model for floods could be suspect, especially if the statistical properties of water flow in the river are nonstationary, for example, because of land-use changes in the river basin or long-term climate change. Assessment of the probability of a volcanic eruption at Yucca Mountain depends both on information about nearby volcanic eruptions over the past several million years and assumptions about the geological processes that create such eruptions (Nuclear Waste Technical Review Board, 1995). These assessments and assumptions are similarly subject to epistemic uncertainty.

In the case of the 1-in-1,000,000 lifetime cancer risk associated with a chemical exposure, such an estimate is often based primarily on indirect evidence and scientific models for exposure, dose, and toxicity. Such models are subject to uncertainty and errors in both their conceptual formulation and the values they estimate for a range of variables affecting how the chemical is transported and transformed in the environment and how the proportion of it that reaches human beings operates in the body. Since the estimated probabilities of cancer are usually well below prevailing incidence rates, the risk estimates are generally not subject to validation or refinement based on epidemiological studies. Thus, barring marked advances in understanding of chemical fate and transport in the environment and of carcinogenesis in humans, full resolution of these uncertainties is unlikely in the near future. Of course, research into individual components of the exposure-dose-toxicity process can help resolve portions of this uncertainty.

Significant advances have been made in recent years in the development of analytical methods for evaluating, characterizing, and presenting uncertainty and for analyzing its components, and well-documented guidance for conducting an uncertainty analysis is available (e.g., Raiffa, 1968; Cox and Baybutt, 1981; Kahneman, Slovic, and Tversky, 1982; Howard and Matheson, 1984; Beck, 1987; Iman and Helton, 1988; Clemen, 1990; Finkel, 1990; Morgan and Henrion, 1990: National Research Council, 1994a). We do not repeat this technical guidance, or recommend specific approaches for uncertainty analysis. Rather, we focus on the role of uncertainty in risk characterization and the role that uncertainty analysis can play as part of an effective iterative process for assessing, deliberating, and understanding risks. In describing this role, we note the critical importance of social, cultural, and institutional factors in determining how uncertainties are considered, addressed, or ignored in the tasks that support risk characterization.

Uncertainties that Matter

Perhaps the most important need is to identify and focus on uncertainties that matter to understanding risk situations and making decisions about them. To accomplish this task, the general approach of decision analysis is helpful. Analysts identify the full set of options for addressing the risk, including options that may extend beyond an initial or limited set of technical fixes or regulatory responses. They then assess the potential impact of each option on the risk problem, using the appropriate natural and social science studies and models. The important uncertainties are those that create important differences in the assessed outcomes and may therefore affect preferences among the available decision options.

Because risk characterization requires providing information about the full set of factors of concern to the interested parties, it must address uncertainty not only about the physical and biological impacts of the risk, but also about the social and political factors inherent to the risk. If social or equity factors matter significantly to the decision, then they deserve at least as careful attention in an uncertainty analysis as do the technical factors, chemical transport properties, dose-response parameters, and so forth.

Another important source of uncertainty lies in the choice of ways to estimate risks and make decisions. The choice of a deliberative process may affect decisions and the ultimate risks in an indeterminate way. It is difficult to predict public reactions to the release of data that are alarming, but of questionable validity: Will it increase or decrease self-protective action? Will it complicate problem resolution or make it easier? Such questions also reveal indeterminacy. When the decision process itself adds uncertainty to risk estimation, efforts to understand and study these process factors and the uncertainties they bring are important to advancing risk characterization.

Purposes

The analysis of uncertainty should elucidate the current state of knowledge and prospects for improving it. As noted by North (1995:278), "Perhaps the most important aspect is not the probability number, but the evidence and reasoning it summarizes." As part of an open, iterative, and broadly based analytic-deliberative process, uncertainty analysis can inform all the parties of what is known, what is not known, and the weight of evidence for what is only partially understood. Describing the uncertainty in the current state of information does not in and of itself represent or imply an advancement in that state; it does, however, help clarify what can be known and perhaps help identify directions for future research and data collection efforts.

As part of the analysis of uncertainty, explicit efforts should be made to identify the activities and resource allocations most likely to yield significant reductions in the uncertainties that matter. Again, these uncertainties may involve the technical-physical components of the risk problem, the social-legal-ethical dimensions, or elements of the evolving processes of risk analysis and decision making. New information will not always reduce uncertainty—it may sometimes provide the knowledge and insight necessary to recognize that the problem is more complex and uncertain than previously recognized. But this too is enlightening, providing an improved understanding of the state of the knowledge pertinent to the risk problem. Such new information should be encouraged, even if it threatens to make the risk problem less tractable. The goal is to provide a comprehensive summary of available and relevant knowledge as the basis for a decision.

Uncertainty analysis also involves assessing the potential for uncertainty to be reduced, which may have important implications for the choice among decision alternatives. Formal value-of-information analysis provides a set of useful techniques for assessing these implications. These techniques involve estimating how risks would change with new information, such as additional experimental results, before that information exists. For example, the artificial sweetener saccharin was considered to pose a cancer risk to humans on the basis of observations of bladder cancer in rats. Additional research during the past 20 years has yielded results that suggest that the physiological conditions under which exposure to sodium saccharin causes bladder tumors in rats may not apply to humans (Cohen and Ellwein, 1995a,b), thus calling into question previous risk estimates for humans. Suppose similar studies could be carried out on other chemicals that are regarded as carcinogens on the basis of animal tests. Even though it is difficult to get definitive results, research to evaluate chemicals that are believed to be carcinogens based on animal tests can be worthwhile when regulatory costs are high. The potential improvement in regulatory decisions, in terms of costs avoided and lives saved, from the study results might have very high value compared with the cost of such studies (North et al., 1994; North, 1995).

Value-of-information methods address whether potential reductions in uncertainty would make a difference in the decision; they suggest priorities among reducible uncertainties on the basis of how much difference the expected reduction might make. They have been useful in helping to identify the value of research and data collection for a number of environmental and related risk issues (e.g., Raiffa, 1968; Howard, Matheson, and North, 1972; Finkel and Evans, 1987; Reichard and Evans, 1989; Clemen,

1990; Freeze et al. 1990; James and Freeze, 1993; Taylor et al. 1993; James and Gorelick, 1994; Dakins et al. 1994; North, 1995). Value-of-information analysis can be of considerable use in the analytic-deliberative process. We emphasize, however, that determining whether a reduced uncertainty would make a difference in a decision often requires deliberation as well as analysis. Different participants in the decision process may not agree on how to interpret new information or on the appropriate criteria for making or revising risk decisions.

Limits

Considerable research highlights the difficulties that experts and non-experts alike have in making scientific judgments related to risk and probability estimation (e.g., Kahneman and Tversky, 1972, 1973; Lichtenstein and Fischhoff, 1977; Kahneman et al. 1982; Freudenburg, 1988; Morgan and Henrion, 1990; Clarke, 1993; Tversky and Kochler, 1994). These difficulties are minimized when the judgment is easy, when there is a clear criterion of accurate judgment, and when those making the judgment have frequent feedback that gives them empirical knowledge about how accurate their judgments are (Fischhoff, 1989). Some risk-related judgments have these qualities—judgments about the frequency of highway accident fatalities may be an example—but many of the most controversial risk judgments do not. Indeed, the biases, imprecision, and overconfidence often associated with expert evaluations of risk provide much of the impetus for conducting an uncertainty analysis. If point estimates of risk are likely to contain significant errors, then explicit evaluation of uncertainty is needed to ensure consideration of the possible sources, magnitude, and implications of these errors. However, just as scientific judgments concerning point estimates are often tenuous and susceptible to overconfidence, so too are characterizations of the uncertainty in these estimates. Formal uncertainty analysis should not be conducted or presented as a final, full, and all-enlightening explication of the risk problem. This is especially true when expertise in risk and uncertainty analysis is unevenly distributed among different parties in an adversarial setting, and formal analysis serves as a tool (whether intentionally or not) to limit participation in and control of the debate. Rather, uncertainty analysis should be recognized as an often helpful technique that, for some problems, can provide insights in support of risk characterization.

A number of findings about the psychology of judgment under uncertainty have implications for the ability of experts both to develop risk estimates and to describe their associated uncertainty (Kahneman, Slovic, and Tversky, 1982). Important among these are the following:

• *Availability*: People (including experts) tend to assign greater probability to events to which they are frequently exposed, e.g., in the news media, scientific literature, or discussion among friends or colleagues, and that are thus easy for them to imagine or recall through mental examples. The "availability" of an event to memory or imagination may not be correlated with the actual probability of the event occurring (Lichtenstein et al., 1978). Indeed, mention in the news media or the scientific literature may occur because the event is rare and unusual. Availability may be one reason that people greatly overestimate the frequency of homicide relative to suicide or the risk of death from accidents relative to the risk of death from diseases (Lichtenstein et al., 1978).

• *Anchoring and adjustment*: People's estimates of uncertain values are influenced by an initial reference value, which may be based on only speculative or illustrative information presented as part of an initial problem formulation, from which they make adjustments on the basis of additional information. Moreover, the adjustment is often insufficient, so that the overall probability assessment is unduly weighted toward the initial anchor value. For example, strong anchoring effects were obtained by Lichtenstein et al. (1978), who had two groups of respondents estimate the frequency of death in the United States from various causes. One group was given the death rate from accidental electrocution (1,000 per year) as a standard of comparison. The second group was given the death rate from motor-vehicle accidents (50,000 per year) as a standard. The second group gave uniformly higher estimates than the first group for all other hazards.

• *Representativeness*: People judge an event by reference to others that resemble it, even if the resemblance carries little or no relevant information. Information that is available or provided on the occurrence of one supposedly representative event can cause analysts to ignore or undervalue large amounts of relevant information. Thus, representativeness has been attributed as the cause of many shortcomings or biases in "statistical thinking," such as failure to appreciate the difference in reliability between small and large samples of data and failure to make one's predictions of future events sufficiently dependent on the overall population mean rather than a few events presumed to be typical.

• *Belief in "law of small numbers" and disqualification*: Many scientists believe small samples drawn from a population to be more representative of the population than is justified on the basis of standard statistical sampling theory. Accordingly, a little evidence can unduly influence the probability assessment. However, people also tend to "disqualify"—that is, discount or neglect—information that contradicts strongly held convictions.

• *Overconfidence*: As a result of these heuristics, many experts overestimate the probability that their answers to technical questions are correct, including probability estimates for risk problems, especially when the questions or problems are difficult and complex.

While these cognitive tendencies are now widely recognized, and techniques have been developed to attempt to address them as part of expert evaluation and elicitation methods (see Spetzler and Stael von Holstein, 1975; Wallsten and Budescu, 1983; Morgan and Henrion, 1990), they provide an important caution. (For statistical models that can be used to account for errors or misrepresentation in probability elicitation and assessment, see Chaloner, 1996; Dickey, 1980; Genest and Schervish, 1985; Kadane et al., 1980; Wolpert, 1989.) A healthy dose of skepticism and humility is appropriate in interpreting any summary of information on risk and uncertainty.

When conducting uncertainty analysis, other cautions and reality checks are in order. First, results of analysis can be very sensitive to assigned probabilities and uncertainties, especially when the estimates involve rare, low-probability events. Freudenburg (1988) demonstrates this for the case of a hypothetical low-probability event that usually presents risk a of 1 in 1 million.[1]

The ability to deal with ignorance and surprise—unforeseen or unforeseeable circumstances—is inherently limited in an uncertainty analysis. Unfortunately, experience shows that it is often these unknown circumstances and surprise events that shake risk analyses and topple expectations, rather than the factors (important though they might be) that have been recognized and incorporated in formal analyses. Examples include the surprising combinations of improbable events that led to the 1979 accident at the Three Mile Island nuclear power plant and an earlier accident at the nuclear power plant at Browns Ferry, Alabama.

Uncertainty analysis should also avoid the temptation to view the evaluation and simulation results that some techniques of uncertainty analysis generate as the equivalent of field and laboratory studies and data. As noted by Morgan et al. (1984:214-215):

> . . . analytical techniques [for uncertainty analysis] . . . are not a substitute for scientific research. They do, however, produce very technical-

---

[1]Freudenburg's calculation is as follows: Assume that 10 percent of the time, the event has a probability of 1 in 1 billion, that 10 percent of the time, the probability is 1 in 1 thousand, and that the remaining 80 percent of the time it is 1 in 1 million. The overall risk is $(0.1 \times 10^{-9} + 0.8 \times 10^{-6} + 0.1 \times 10^{-3})$, which equals .0001008001, or slightly more than 1 in 10 thousand—a much larger number than the most likely value.

looking results and it is usually faster and cheaper to go ask a group of experts what they think than it is to sponsor the research that is needed to learn the true answers. In agencies pressed for quick decisions, operating on short time constants, and staffed by many people who do not have technical backgrounds, there is a risk that these techniques will inadvertently become a substitute for science.

Although careful, well-focused, and appropriately modest applications of uncertainty analysis should be helpful for many problems, there are situations in which there is simply no need for formal methods of this type. This may be the case in simple, repetitive, and highly institutionalized settings where the administrative need for consistency and standardized, "bright-line" decision rules may outweigh the need to characterize the uncertainty of the consequences of a particular decision (though an occasional review to assess the ongoing performance and uncertainty of the overall decision-making process is still in order). Also, formal uncertainty analysis may not help if the uncertainty in the fundamental understanding of the basic processes that drive the risk, or of whether the risk is even present at all, is so large that a quantitative estimate can only lead to obfuscation. An example is the possibility that global emissions of greenhouse gases could lead to a drastic change of state such as shutting off the North Atlantic Ocean circulation pattern (the Gulf Stream), leading to a drastically colder climate in Northern Europe. Both the probability of occurrence of such an event and the range of possible consequences should it occur are extremely difficult to characterize. In such cases, identification of important issues and perhaps some selected analysis of scenarios (without assigning probabilities to these scenarios), is the best that can be accomplished.

Social Context

Various social, cultural, and institutional factors affect how people recognize and use information on uncertainty. Understanding depends not only on the inherent features of a risk, or even the experience and expertise of the analyst attempting to characterize it, but also on the social context of the risk analysis and the associated deliberative process (e.g., Brown, 1989; Jasanoff, 1987a, 1987b, 1991; MacKenzie, 1990; Michael, 1992; Shapin, 1994; Thompson, Ellis, and Wildovsky., 1990; Wynne, 1980, 1987, 1995). These factors affect the way information about uncertainty is created and utilized in evaluating risks and the degree to which analysts acknowledge uncertainty.

Cultural and social factors affect whether or not uncertainty is openly recognized in risk characterizations. In many legal settings, for instance, the proceedings are expected to produce a sharp boundary between truth and belief through "fact-finding." Scientific and social institutions that must maintain trust and authority as the interpreters of scientific truth and that must support a clear legal finding can often display a purposeful ignorance or pushing aside of information on uncertainty. Suppression of uncertainty can also operate through the group processes of consensus building, for example, during the deliberations of scientific advisory panels and expert bodies, even when there is no legal mandate for a single outcome or recommendation.

When the stakes in a decision are high, accuracy or inaccuracy in science may be accentuated by participants for their own purposes. For example, in the early 1980s a debate over acceptable levels of polychlorinated biphenyls in the ground around leaking transformers (for example, on electric power poles) highlighted existing uncertainty about the health risks. Environmentalists argued that cleanup should be to the level of detection (about 5 parts per million [ppm], at that time), while several industry groups argued that a 50 ppm cleanup level should be considered safe. Because of the uncertainty in available health studies, both positions received scientific support, but neither could prevail. Eventually both sides concluded that achieving an acceptable cleanup policy would yield more benefits than an unending argument about health effects. They reached a compromise that included a 25 ppm cleanup standard and jointly persuaded the U.S. Environmental Protection Agency and Congress to implement their compromise; the uncertainty ceased to matter to the parties (Bannerman, 1987; Warren, 1987).

The perception of uncertainty tends to vary with closeness to the problem—those very close to or far from a problem often acknowledge the greatest uncertainty, while those with some partial knowledge tend to consider their understanding to be more definitive, suggesting a "trough of uncertainty" (MacKenzie, 1990), or, perhaps, that a little knowledge can be dangerous to understanding. Perceptions of uncertainty can also be greatly influenced by the cultural and social context of the perceivers' experiences and their roles in relation to the risk problem. Judgments of the uncertainty of scientific information often reflect the trust and reliability placed in the institutions that have generated the information. For most people, an investment of time and energy to understand scientific information and its uncertainty is only worthwhile when they may be affected or when such information is relevant to decisions over which they have the power and agency to act and make a difference. In some cases, interested parties may even seek technical ignorance when such behavior is socially beneficial or appropriate, such as when knowledge can impart responsibility or liability for a risk or when pursuit of such knowledge can signal mistrust in actors or social arrangements upon which they depend for support or protection. Scientific theory and approaches assume that more information and less uncertainty is always preferred, but this may not be the case in some cultural and social situations.

Summary and Implications

Uncertainty is a critical dimension in the characterization of risk. Participants in decisions need to consider not only its magnitude, but also its sources and character—whether it is due to inherent randomness or to lack of knowledge and whether it is recognized and quantifiable, recognized and indeterminate, or perhaps unrecognized. Uncertainty is best examined in the context of a decision, focusing on the uncertainties that matter most to the ongoing deliberation and decision processes. These uncertainties may involve the physical and technical aspects of the risk, the social and economic dimensions of the risk, or political or behavioral factors that influence the evolution of the risk and associated uncertainty. By focusing on these factors in a decision-analytic context, uncertainty analysis can enlighten decision participants, help counter the cognitive biases that affect expert judgment on risk, and help set priorities for further information gathering efforts.

Uncertainty analysis should be conducted with care and in conjunction with deliberation. Although uncertainty analysis can be a useful tool for more informative characterization of risk, it has limitations. It cannot address the truly unexpected—the risks that were never considered in a risk analysis but that arise with unknown frequency in real events. It can at times be misleading, and in certain cases, may have no appropriate role at all. Moreover, cognitive biases can affect judgments about uncertainty as well as about risk. Uncertainty analysis and its users should remain aware of the fact that both the analysis and people's interpretations of it can be strongly affected by the social, cultural, and institutional context of the decision setting and the formal or perceived role of the various participants, which can exert pressure toward perceiving more or less uncertainty, or different kinds of uncertainty, than would otherwise be recognized.

CONCLUSIONS

Analytic techniques can be used for several aspects of risk characterization. Most familiar among these uses is to estimate the likelihood of particular adverse outcomes. In addition, they are often used to reduce inherently multidimensional risks to a single dimension so as to facilitate decision making, and to characterize the uncertainties surrounding estimates of adverse outcomes. Much insight can be gained from applying analytical techniques to these purposes, and there are strong practical reasons for decision makers to seek standardized, replicable, and defensible analytic procedures. However, there are important pitfalls associated with overreliance on analysis. Analysis conducted to simplify the multidimensionality of risk or to make sense of uncertainty can be misleading or inappropriate, can create more confusion that it removes, and can even exacerbate the conflicts it may have been undertaken to reduce. Because of the power of formal analytical techniques to shape understanding, decisions about using them for these purposes and about interpreting their results should not be left to analysts alone, but should be made as part of an appropriately broad-based analytic-deliberative process. Used in this manner, proper analysis can enlighten both scientific understanding and the goals of effective risk decision making.

5

Integrating Analysis and Deliberation

As detailed in the previous two chapters, well-informed risk decisions depend on both analytic and deliberative processes. Although this is widely recognized, there are some common misunderstandings about the roles of the two processes. One is that analysis and deliberation proceed in a sequence—for example, that deliberation is used to make decisions after risk analysis has been completed. Another is the idea that knowledge comes only from analysis and that the role of deliberation is solely in making decisions. In fact, however, both analysis and deliberation have contributed and can contribute to each step of the process leading to risk decisions. Both are processes for increasing understanding about existing phenomena and estimating future conditions. Both are also ways of informing, constructing, and testing judgments about the validity of evidence and the appropriateness of decisions—not only substantive ones, but also the many procedural and methodological ones that lead to effective risk characterization. Thus, both analysis and deliberation are useful in every step leading to risk characterization, and participants in risk decisions are likely to be better informed if the two processes are combined in appropriate ways. Finding the appropriate balance and interaction for each step and each decision context is the challenge.

We cannot prescribe a standard procedure for meeting this challenge. Rather, organizations need to be creative and flexible when devising processes to engage and inform the participants in decisions. We recognize that one of the greatest difficulties of doing risk characterization is tailoring an approach that is effective, efficient, and appropriate for the specific risk and the social and institutional conditions surrounding the risk decision. This requires judgment on the part of the responsible officials. This chapter offers some guidance and experience on how to integrate analysis and deliberation at each major step, to inform those judgments. It also addresses one of the key practical problems in risk analysis and characterization—the tension between the desire for more analysis and deliberation and the need to reach closure. Chapter 6 offers guidance on how to match the analytic-deliberative process to the needs of specific decisions.

We emphasize at the outset that the most extensive forms of analytic-deliberative process are appropriate in only a relative few instances. These instances, however, have an importance disproportionate to their number, and it is not always evident in advance when a risk characterization will require extensive deliberation, integrated with analysis. For example, it is appropriate to develop standard procedures that can be used to characterize the risk associated with large classes of routine and narrow-impact decisions, and it may be appropriate to do the same for some classes of repeated decisions with wider impact, such as siting power plants (see Chapter 6). But such choices to routinize should themselves be matters of broad deliberation to ensure that any resulting routines will meet the needs of public officials for information and of the interested and affected parties for information and participation.

The appropriate breadth of participation in an analytic-deliberative process also depends on the situation. One important factor is whether particular parties are likely to be affected by a decision: the organization responsible for the risk characterization should consider including potentially affected parties even if they do not yet realize they may be affected. Another factor may be the level of trust the parties have in the commitment and ability of the technical experts and the decision-making organizations to protect them. It may be, for example, that the U.S. airline industry and its regulators enjoy greater public trust at present than the U.S. chemical industry and its regulators, so that broad inclusion may be a greater practical necessity for characterizing the effects of technical decisions about chemical risks than about aircraft risks. Levels of trust change, of course, and inappropriate decisions to limit participation sometimes contribute to loss of trust.

This chapter illustrates integration of analysis and deliberation with examples linked to particular steps of the process that informs risk decisions. (See Figure 1-2, page 28, for a representation of the steps in the risk decision process.) The examples are offered as illustrations only, not as specific prescriptions for future use. In many of these examples, careful evaluations were not done; moreover, there are no clear definitions of success available for the parts of the analytic-deliberative process. In addition, some of the efforts we describe are themselves controversial, a fact that illustrates the difficulty of designing an effective analytical-deliberative process for informing potentially contentious risk decisions. We nevertheless consider the examples instructive for practitioners who are trying to organize risk characterization in ways that take advantage of the strengths of both analysis and deliberation.

Although the examples are specific to particular steps of the process, it is important for the entire process to attain coherence, in the sense that the participants in the decision understand the procedures and find them sensible. Examples in this chapter show that coherence can be achieved in various ways. The Florida Power Corporation (also see Appendix A) used repetitions of a similar procedure to move from a long list of options to a recommended site for its proposed power plant; the three-step process (Renn et al., 1993) uses different procedures at different steps, with organized interests taking the lead in the first step, technical experts in the second, and randomly selected citizens in the third. Although neither of these approaches has been used often enough to recommend it confidently for adoption, each may nevertheless be useful as a source of ideas for officials charged with organizing risk characterization efforts.

PROBLEM FORMULATION

Problem formulation can be contentious because the way a risk problem is framed partially determines the way risks are analyzed and understood, thus affecting decisions (see, e.g., Vaughan and Seifert, 1992). For example, a problem having to do with waste disposal might be framed as one of too much waste, too little recycling, or too little disposal capacity. Such problem framings may be linked to interest positions. In this case, for instance, waste haulers usually prefer to solve the problem of inadequate disposal capacity rather than solving the problem of excessive waste generation. In practice, however, it is not always easy to determine how a problem formulation affects one's interests. What once seemed to be the most desirable solution may not seem so after the affected parties have had an opportunity to present their knowledge and perspectives.

Both analysis and deliberation can aid in problem formulation. Analysis has often provided the first news that a hazard may exist, and it can supply much useful information about the nature of the hazard, as well as about the feasibility and likely consequences of different ways of eliminating or mitigating it. Deliberation that includes interested and affected parties sometimes elicits ways the problem could be redefined, as well as insights about which problem definitions are likely to be widely accepted. The following example, from a regulatory negotiation on disinfectant by-products sponsored by the U.S. Environmental Protection

Agency (EPA), illustrates how deliberation among the parties, informed by analysis, was used to help frame the risk decision problem. (As the more detailed case description in Appendix A shows, this regulatory negotiation also combined analysis and deliberation in accomplishing other tasks.)

The regulatory negotiation on disinfectant by-products—which are drinking water contaminants—involved representatives of major "stakeholder" groups in a process aimed at advising the EPA on proposing rules for regulation. Disinfectants, mainly chlorine, are used in drinking water to kill microbial pathogens; they react chemically with naturally occurring organic compounds in the water to produce by-products that are carcinogenic.

The negotiating committee soon uncovered disagreement about the nature of the decision problem. A previous EPA report had framed the problem as a risk-risk tradeoff: reduction of risk from the by-products was linked to an increase in microbial risk. Thus, a change to nonchlorine disinfectants would reduce the risk from certain by-products, but it might increase risks from pathogens or other disinfectant byproducts whose effects were not well studied or understood. Although most members of the negotiating committee agreed that some type of rule was needed, not all were ready to accept EPA's definition of the problem as a risk-risk tradeoff that could only be resolved by setting and enforcing "maximum contaminant levels" for disinfectant by-products. Some believed that a reduction of by-product precursors in the water would reduce the need for disinfectants, thereby side-stepping the risk-risk tradeoff. They wanted the committee to consider a pollution prevention approach aimed at eliminating the organic precursors of the by-products—naturally occurring humic and fulvic acids that react with disinfectants to produce dangerous by-products. Theoretically, the problem would disappear if precursors could be removed before the water had to be treated. Precursors can be reduced though watershed protection measures (which prevent these precursors from ever becoming highly concentrated in the water reservoir) or by pretreatment filtration (which removes precursors before the disinfectant is added). The idea that disinfectant by-product precursors could be reduced through improved watershed management or the addition of a removal technology was attractive to some. This led to proposals to investigate enhancing the existing Surface Water Treatment Rule as a means of controlling disinfectant by-product precursors.

The negotiating committee realized that more data and analysis might help it decide whether a reformulation of the problem could yield more acceptable and protective solutions. The committee called for assistance from the technical advisory committee, a group of scientific and technical experts that had been appointed at the start of the process to support the analytical needs of the negotiating committee. The technical advisory committee organized a technical workshop to inform negotiating committee members on the range of scientific opinions about health risks, treatment technologies, costs, and modeling efforts. Twenty-three nationally recognized experts on drinking water treatment gave presentations and participated in panel discussions for the benefit of the negotiating committee. As questions arose during the negotiations, these experts gave additional presentations or testimony.

The analysis did not lead the committee to accept a single problem formulation, largely because both formulations were supportable. While watershed protection was attractive, new data suggested that it would not be sufficient for controlling disinfectant by-products. Some form of contaminant rule would also be necessary. The negotiating committee was able to proceed without consensus on the problem because, by using deliberative methods, it was able to agree on a set of criteria for an acceptable solution. In professionally facilitated open discussions, the committee produced a list of value objectives to be considered in the decision-making process. This included such items as protection of human health, protection of environmental equity, sensitivity to needs of susceptible populations, consistency with EPA rules, and affordability. These criteria were later used as a basis for discussing proposed rules. In the end, the committee proposed both a pollution prevention rule and a maximum contaminant level rule—an outcome not anticipated before the analytic-deliberative process began.

PROCESS DESIGN

Process design determines who participates, what their roles are, how analysis will be organized and used, and how procedural rules can be changed. One of the most important goals of process design is to devise procedures that are acceptable to the interested and affected parties. Obtaining agreement on a decision process at the outset from those who will be affected by the decision can significantly affect the acceptability of the outcome (Crowfoot and Wollendeck, 1990). Analysis and deliberation can complement each other in achieving two of the main objectives of process design: determining who should be involved in risk characterization and planning for the appropriate use of analytical techniques.

Analysis can help choose participants in at least two ways. First, it can help identify the affected parties. Analysis of exposure pathways, dose-response relationships, subpopulation vulnerabilities, the distribution of economic and social impacts, and the like can identify parties who should be involved, even if they did not know themselves to be affected. Analysis of the same factors can also provide reasonable tests of parties' claims to be affected. Second, analysis of legal obligations can clarify the fundamental requirements of public participation—who is required by law to participate and who has standing to challenge decisions about participation.

Deliberative techniques are also essential for choosing participants. They can be used to address such questions as the following: How much representation should each interested or affected party have in the group performing a particular task relevant to risk characterization? Who should make this decision? When an affected party has no obvious representative (e.g., future generations), how should its interests be represented? What kinds of expertise should be included in the group responsible for each task? A deliberative tool such as a citizen advisory committee might be used to agree on answers to such questions and to select principles for choosing participants for each subsequent phase.

A broadly based process that includes analytic specialists and others can improve planning for the appropriate use of analysis to support a risk characterization. Among the most important judgments is the one about which aspects of a risk situation to analyze. The California Comparative Risk Project, which attempted to rank issues for the purpose of setting statewide policy priorities, illustrates a process design that provided for feedback between analysis and deliberation to inform choices about what to analyze. The California Environmental Protection Agency's initial process design, resulting from its diagnosis of the task, sharply separated analysis and deliberation by assigning them to different groups of committees that would interact only toward the end of the process (see Appendix A). When the project began, some participants criticized this process design on the ground that without early input from nontechnical people, the technical committees might fail to address important issues, such as equity in the distribution of risks. The California agency responded by redesigning the process to allow for more crossfertilization between the technical committees and those emphasizing social and economic concerns. An immediate result was increased analytical attention to social equity outcomes. A longer term result was more open debate about the proper place of such considerations in risk analysis in California, a debate that resulted in the governor's distancing himself from the study's findings and recommendations.

The Future Site Uses Working Group organized at the U.S. Department of Energy's (DOE) Hanford site in Washington State provides another example of how analysis and deliberation can be combined in process design (see Appendix A). The DOE decided to seek widespread participation in planning its environmental impact assessment for the site. Deliberation among DOE, EPA, the Oregon and Washington state governments, and county and tribal governments of the region produced a list of potential participants for a broadly based working group that would help in this planning. An early analytic activity involved a set of interviews with prospective members of the group to get names of other possible members and ideas for process design. The main deliberative tool was the working group itself. The group, which consisted of representatives of various interested and affected parties, decided that its main task was to identify alternative scenarios for cleanup and future site use, and focused on how these would be connected. Its deliberations included its own meetings, outside review at a series of public meetings, and consultations between its members and their constituencies. The result was that the group specified a set of outcomes to be addressed in subsequent analyses (the environmental impact assessment), and identified outcomes of concern in relationship to particular future uses. Thus, the group's deliberation fed into the environmental impact assessment by suggesting directions the analysis should take.

SELECTION OF OPTIONS AND OUTCOMES

The discussion of problem formulation has already suggested ways in which analysis and deliberation together can help in choosing which actions to consider. In the disinfectant by-products negotiation, defining the problem as a risk-risk tradeoff implied analyzing various limits on maximum contaminant levels, whereas defining the problem in terms of controlling disinfectant by-product precursors suggested different options, including watershed protection and filtration.

Analysis and deliberation can also work together to generate options when the problem is well defined. Numerous examples come from the siting of hazardous facilities. Initially, a very wide area may be open for consideration. Analysis can reduce the options by use of exclusionary criteria that may be legal in nature (e.g., covering National Parks, wetlands, and other protected areas) or physical (e.g., based on the size, geology, or hydrogeology of the site). Deliberative groups, such as advisory committees and citizens' panels, can help develop exclusionary criteria that are not legally or technologically mandated. They allow consideration of diverse, sometimes competing decision criteria, many of them associated with different interests (e.g., industrial versus residential development), values and principles (e.g., one group wants to protect agricultural heritage and another defends property owners' rights to decide), or tendencies to be risk averse under uncertainty (e.g., people may differ, given the same information, about whether a 100-foot setback from a water supply well is far enough to protect drinking water quality).

In 1989, when the Florida Power Corporation (FPC) sought a site for a new coal-fired power generation station, it used a phased site selection process that combined analytical tools and deliberative processes to find a site in a large search area that included the entire state of Florida and the southern portions of Georgia and Alabama (see Appendix A). An initial set of exclusionary criteria developed by FPC staff and consultants reduced the search to 172 potential areas. Four subsequent phases of the process excluded sites on the basis of additional criteria that emerged from two related deliberations: one involved FPC staff; the other involved an environmental advisory group composed of community leaders in the search region, including leaders of environmental and business groups and past officials of local and state governments. These deliberations generated additional exclusionary criteria and assigned them weights. These outputs were shown to both groups, and the FPC group revised its judgments to move them closer to those of the advisory group. The consulting firm then applied the exclusionary criteria and weights to information about the sites and reduced the number of sites, in steps, from 172 to 61, to 21, to 6, and finally to 1 preferred site and 2 alternates.

Analytical and deliberative processes can also be combined to help decide which outcomes to examine. The California Comparative Risk Project and the Hanford Future Site Uses Working Group both illustrate the role of deliberation in making such decisions. A hypothetical example can further clarify this process. Suppose a state agency is considering a rule to regulate exposure to radon from water wells with high radon concentrations. It is considering whether to require owners of seriously affected water wells to install radon-removal systems before selling their properties. Many outcomes of radon exposure might be serious enough to affect the decision: health effects on water users, impacts on property value, costs of implementation to homeowners, health and safety risks of installation, the socioeconomic and racial equity of the proposed rule, compliance considerations, and undoubtedly others. Which of these deserve careful analysis? It may not be reasonable, responsible, or necessary to study all of them.

Analysis can help by giving a preliminary indication of the magnitude of particular outcomes. Analysis of the distribution of exposure to water from high-radon wells along economic, racial, or geographic lines would indicate whether or not any of these are potential concerns. Such an analysis would use statistical, epidemiological, or economic techniques. Analysis using public opinion surveys, focus groups, or interviews may help with other issues, such as compliance and the potential effects of a rule on property values.

Deliberation involving interested and affected parties may help by identifying previously unrecognized possible adverse outcomes that may require further analysis. For example, farmers may ask about radon uptake and damage to farm animals and farm families. Many farms use only well water. Deliberation may also help in defining the criteria to use for choosing which of the potentially significant outcomes deserve significant analytic attention. A state agency might use a variety of deliberative strategies, including public meetings in regions of the state where radon in water is a problem; formal public workshops, including educational sessions and group discussions; or a broadly representative statewide advisory committee.

An organization's final decision on allocation of research efforts needs to build on both deliberation and analysis so that it reflects both informed public opinion and the best expert knowledge. It needs to take into account both the information desired by affected parties and the information that public officials and affected parties will need in the future, given the best understanding of the risks.

INFORMATION GATHERING AND INTERPRETATION

The role of analysis in providing information for informing risk decisions is well known and has been the subject of many volumes of research. Here we discuss the role of deliberative processes, particularly those that include interested and affected parties, and their integration with analysis. Integration can occur in two key ways: deliberation can frame analysis and deliberation can interpret analysis.

Broadly based deliberative processes can raise questions, suggest alternative ways to interpret or frame issues, generate hypotheses, or provide data as input to an analysis of a risk situation. For example, individuals with specialized knowledge about actual operations in organizations engaged in hazardous activities (e.g., nuclear power plant operators, air traffic controllers, coal miners) can identify variables to include in exposure analyses. Interested and affected parties can also provide essential information about what must be analyzed if a risk characterization is to meet those parties' needs for understanding.

In the Florida Power Corporation's siting process (see Appendix A), the broadly based environmental advisory group identified an outcome condition that did not appear on the company experts' initial list, but that became pivotal to the final choice. Their key concern was with what came to be called "proximity to disturbed areas"—that any power plant be sited not only far from people, but also close to areas that had already been environmentally disturbed, so as to preserve pristine areas from development. All the possible sites were then rated on proximity to disturbed areas, and that information strongly influenced the outcome. In the South Florida ecosystem management case (see Appendix A), scientists who had contact with interested and affected parties expressed what they understood to be the parties' informational needs and concerns in the deliberation about how to conduct the analysis, with the result that the analysis took into account not only ecological effects, but also various social and economic effects. Similarly, the Hanford Future Site Uses Working Group deliberated about what issues to address in an environmental impact analysis.

Broadly based, scientifically informed deliberations are also useful for considering the meaning of available information about a risk. In fact, such deliberations are often critical to achieving an understanding that will make a risk characterization credible to its various users and audiences. The reports from the South Florida ecosystem restoration project and the group Delphi process we describe in the next section reflect such deliberations among groups of scientific and technical experts who bring a variety of perspectives to an issue, consider a body of knowledge, and try to arrive at agreement on what it means for decision makers. Citizens' juries (see Appendix B) and the citizens' panels we describe below are broad nonspecialist groups that arrive at interpretations of technical knowledge by a deliberative process that is informed by testimony from specialists.

SYNTHESIS OF INFORMATION

Critical to the success of risk characterization is the task of synthesizing the state of knowledge about the risk situation. Synthesis may involve a final written document, but it can also take other forms, such as an oral presentation, an interview, or an expert workshop. In the regulatory negotiation over disinfectant by-products, oral presentations to the negotiating committee were supplemented with written reports. The test of success is how well the synthesis meets the needs of the range of participants in the decision.

Various analytic tools can be used to summarize information about risks. These include ordinary statistical techniques, techniques for estimating and representing uncertainty, and mathematical models of risk situations that organize the best available data into forms usable for policy analysis. These techniques are not usually integrated with broadly based deliberation or even made "user friendly" for members of interested and affected groups who lack strong technical backgrounds, but there are ways to do so.

The "three-step model" developed by Renn and his colleagues has been used with some success in Western Europe to structure national policy debates and to inform decisions about siting waste disposal facilities (Renn et al., 1991, 1993). This model illustrates how deliberation and analysis can be coordinated at various phases of a public decision-making process, including synthesis. Each of the model's three steps coordinates analysis and deliberation.

In the first step, project staff interview representatives or key members of interested and affected groups, and sometimes also conduct content analyses of newspapers, to generate a tentative formulation of the problem, a list of decision options, a listing of the values or interests that might be affected by the decision to be made, and a list of outcomes to evaluate. The participants need not agree about which values or outcomes to put on the list: at this stage, any outcome suggested by one of them is included.

In the second step, technical experts assess how each choice option may affect each outcome on the list. For instance, in one application relating to the siting of a landfill, technical experts were asked to assess risks and uncertainties in terms of the adverse outcomes identified as important by the potentially affected groups. Project staff identify the experts by conducting interviews, reviewing the literatures, and taking suggestions from agency staff and interested and affected parties. The experts conduct their work using a face-to-face structured communication technique called a group Delphi, which iterates individual responses and group discussions in an effort to seek consensus and define disagreement (Webler et al., 1991). The group Delphi combines analysis with a deliberation among the experts. The results of the group Delphi are summarized in two ways: in a written report presenting the experts' quantitative estimates of risk and uncertainty, and in video-taped testimonies made by scientists with different views, which show the differences in judgments about how the data should be interpreted are summarized.

In the third step, the videos (or sometimes direct testimony) are presented to panels consisting of representative or random samples of the affected citizens. The citizens' panel or panels take part in a series of working sessions during which they learn about the scientific analysis of the problem from experts, who may not agree with each other. The citizens' panels may also solicit more analysis or conduct their own inquiries. In these sessions, the panelists gain an understanding of the risk situation and, after considering the possible outcomes and the expert judgments about their likelihood given each option, they deliberate and make recommendations about the final decision. The recommendations are given to the responsible agency which, ideally, has made a prior commitment to implement the panel's recommendations if at all feasible.

The three-step model illustrates some ways to integrate analysis and deliberation to summarize the state of knowledge about a risk situation: the group Delphi uses deliberation among technical experts to provide a written synthesis; the videotaped testimony summarizes expert opinion about technical data in a way that allows the citizens' panel to deliberate about what it means; and the panel's active efforts in interviewing experts and soliciting more analysis also uses analysis to inform deliberation. Note that only the group Delphi produces a written report. For the most part, the synthesis of knowledge about risks and uncertainty is not achieved by a quantitative, analytical method but rather by a process that combines analysis and deliberation among experts (in the second step) with scientifically informed deliberation by nonexperts (in the third step). In essence, the citizens' panels are empowered to keep asking the experts to characterize the risks until they are satisfied. Note also that there is no effort to get the panel to agree on the state of knowledge—there is no attempt to present a single authoritative summary and, apparently, no need for one. The panel sees expert opinion, complete with its uncertainty and disagreements, and appears to be able to use this view of the state of knowledge to generate an implicit synthesis that informs its recommendations for action.

ACHIEVING CLOSURE

Since the analytic-deliberative process leading to risk characterization is iterative, no step of it is closed in a permanent sense. By *closure*, we refer to a decision to end, wrap up, or call off an ongoing activity and move on to the next, even if revisiting the present one remains a possibility. Because of real-world deadlines for decisions, whether set by law, budgets, or competing work, it is the responsibility of the organization charged with preparing a risk characterization to determine the point of closure for each step of the analytic-deliberative process. The organization is also in the best position to create mechanisms to promote closure and to set and enforce criteria for closure.

Reaching closure is not a serious problem in many analytic-deliberative processes. It is likely to be most difficult when interests are in strong opposition; when the number of participants is large; and when differences are based on fundamental values, as opposed to interpretations of evidence or motivation. Organizations should anticipate the possibility that the need for closure will come when the participants in a deliberation have not reached consensus. Under these conditions, an organization should consider two reasons to delay closure: to allow all parties a fair chance to hear others and be heard, and to bring to the surface additional information, concerns, and perspectives that will need to be considered if a risk characterization is to address the needs of the decision makers, public officials, and the interested and affected parties.

Organization officials need to take care not to be or appear arbitrary in closing a part of the process. Inappropriately early closure decisions, even if they are legal, may destroy the rapport that the organization has built with interested and affected parties and may result in lost credibility and opposition to the ultimate decision, either in the form of legal challenges or in political arenas. Agencies cannot and should not expect to satisfy all the interested and affected parties, but they, and their missions, can frequently benefit if they are more responsive to the parties' procedural demands than the law requires them to be.

Those who manage an analytic-deliberative process should consider the intent of the participants when making decisions about closure. If the participants appear to be struggling to find a solution that is in the common good, it may be prudent to invest more time or resources in continuing the process. However, as noted in Chapter 3, some parties may sometimes be motivated by hidden agendas. For example, parties on all sides of risk debates have used demands for more analysis or more deliberation in order to delay decisions (Ozawa, 1991; Bingham, 1986). This strategy may advance the agenda of a participant, but at the expense of the broader purposes of the analytic-deliberative process. Strategic delay may also discourage other parties from participating or push organizations to close avenues for meaningful participation.

If the organization anticipates that the process may be prolonged by one or more participants to pursue their own interests, it may be prudent to encourage the participants to adopt constraints, or even to impose constraints or close the deliberation by the authority of the convening organization, once all the information has been elicited and the viewpoints and perspectives aired and adequately discussed.

A government agency or other organization may press for closure by restricting the budget or setting a deadline. This is a common technique used in mediation and regulatory negotiation. It may be attractive to agencies because it attaches a precise dollar figure or timetable to the task, but it can leave the agency vulnerable to charges that the constraints were inappropriate or insincere. Constraints imposed from outside an agency may also help in reaching closure, but some participants may consider them illegitimate. When participants consent to the constraints ahead of time, claims of illegitimacy are more difficult to sustain.

Another way to promote closure is to have the participants in a deliberation adopt a set of procedural rules that can be used in their discussions to reach closure even when substantial disagreements persist. For example, they may decide to follow parliamentary rules and resolve issues by majority vote after a discussion period. Consensus, in which each participant is given a veto, is another possibility that is used in regulatory negotiation. Another is the half veto—when any two participants can veto a decision.

Organizations may also impose constraints on the process and seek informed consent to those constraints (see Shrader-Frechette, 1993b:367).

When the participants agree to constraints that will enable the process to be closed, they essentially commit themselves to accepting the closure when it occurs. Organizations should look first for ways to obtain agreement from the participating parties in advance to constraints that can force an analytic-deliberative task to closure.

Regardless of their content, procedural rules for ending deliberation should be tailored to the needs of the situation and the deliberative body. Different rules can be assigned to different kinds of closure decisions, in the same risk decision process. Such an approach has worked well with citizen advisory committees and other forms of citizen panels (Lynn, 1987a; Renn et al. 1993). It has the advantage that it can generate rules that protect groups from being forced into compromises on issues of fundamental importance to them.

A significant problem with closure can arise when a government agency responsible for characterizing a risk is motivated to postpone the decision (see Graham, 1985; Dwyer, 1990). Officials may then claim, without what others consider adequate reason, that more detailed analysis or deliberation is necessary before taking action. Such a motive may conflict with the needs of the larger society and may be difficult to counteract within the agency itself. This possibility requires alertness on the part of the interested and affected parties and appropriate authorities in the executive, legislative, or judicial branches of government.

CONCLUSION

This chapter and the two preceding ones have detailed an expanded view of the risk decision process that can be the basis for successful risk characterizations. Structuring an analytic-deliberative process for informing a risk decision is not a matter of formal blueprints or step-by-step directions: every step of the process, from identifying possible harm to deciding when to close the last part of an analysis or the last part of a deliberation, involves judgment. The right choices are situation dependent.

To guide their judgments, those who manage risk decisions can rely on a few principles, which we present in Chapter 7. By following these principles, organizations can expect, over time, to improve their ability to design processes that lead to successful risk characterizations. Success in risk characterization, like the rest of the risk decision process, cannot be measured by a checklist or a formula. But we believe success can be measured against several criteria that grow out of our overall framework. We see each of these criteria as a necessary, but not sufficient, condition for satisfactory risk characterization.

- *Getting the science right:* The underlying analysis meets high scientific standards in terms of measurement, analytic methods, data bases used, plausibility of assumptions, and respectfulness of both the magnitude and the character of uncertainty, taking into consideration limitations that may have been placed on the analysis because of the level of effort judged appropriate for informing the decision.
- *Getting the right science:* The analysis has addressed the significant risk-related concerns of public officials and the spectrum of interested and affected parties, such as risks to health, economic well-being, and ecological and social values, with analytic priorities having been set so as to emphasize the issues most relevant to the decision.
- *Getting the right participation:* The analytic-deliberative process has had sufficiently broad participation to ensure that the important, decision-relevant information enters the process, that the important perspectives are considered, and that the parties' legitimate concerns about inclusiveness and openness are met.
- *Getting the participation right:* The analytic-deliberative process satisfies the decision makers and interested and affected parties that it is responsive to their needs—that their information, viewpoints, and concerns have been adequately represented and taken into account; that they have been adequately consulted; and that their participation has been able to affect the way risk problems are defined and understood;
- *Developing an accurate, balanced, and informative synthesis:* The risk characterization presents the state of knowledge, uncertainty, and disagreement about the risk situation to reflect the range of relevant knowledge and perspectives and satisfies the parties to a decision that they have been adequately informed within the limits of available knowledge. An accurate and balanced synthesis treats the limits of scientific knowledge (i.e., the various kinds of uncertainty, indeterminacy, and ignorance) with an appropriate mixture of analytic and deliberative techniques.

These criteria are related and mutually complementary. The defining feature of good risk characterization is that it meets the needs of decision participants. It must therefore be accurate, balanced, and informative. This requires getting the science right and getting the right science. Participation is important to help ask the right questions of the science, check the plausibility of assumptions, and ensure that any synthesis is both balanced and informative. The more likely it is that the science will be criticized on the basis of its underlying assumptions or alleged omissions, the more important participation is likely to be in a risk decision process.

6

Implementing the New Approach

The previous chapters call on organizations that engage in risk analysis and characterization to do things they do not routinely do: combine analysis with deliberation, broaden the range of outcomes potentially subject to analysis, and broaden participation in activities that were previously restricted to analytic experts and a few decision makers. It may seem to some readers that implementing such an approach would be quite impractical because it would require a major increase in the effort made to characterize risks at a time when the responsible organizations are already overloaded and resources are stable or shrinking. We believe, however, that when the effort is appropriately scaled to match the needs of the decision at hand, it does not necessarily require more time and money and that when it does, the potential benefits are likely to outweigh the costs. This chapter discusses the issue of practicality as it affects our approach to risk characterization. It then discusses two keys to making our approach practical: diagnosing risk decision situations in order to match the process to the needs of the situation and building the capability for implementation.

PRACTICALITY

There are legitimate concerns about the practicality of the analytic-deliberative approach to risk characterization presented in Chapters 3, 4, and 5. One concern is with the costs and benefits of expanding the concept of risk analysis to include ecological and social outcomes as well as threats to human health and safety. An expanded domain for risk analysis might require the responsible organizations to hire new kinds of experts and to support new and expensive analyses, some of them relying on new techniques. It is reasonable to ask whether such additional activity can lead to better or more acceptable decisions and whether in a context of restricted budgets, it might displace other, more essential, efforts. Another concern is that bringing interested and affected parties into scientific and technical discussions, such as about which analytic technique to use, might introduce delay and confusion and allow nonscientific concerns to impinge improperly on scientific decisions. A third concern is that adding participants and increasing the number of issues that must be considered can provide many opportunities for any interested or affected party that might benefit from delaying a decision to find excuses for delay.

Such adverse outcomes might indeed result from adopting our approach to risk characterization, and it is certainly true that expense, delay, and confusion already plague risk characterizations. On balance, however, we believe the approach we propose is more likely to mitigate these outcomes than cause them, especially when applied to major decisions with potentially wide impacts.

Consider, for example, the potential for expensive and time-consuming debate about the adequacy of risk analyses. Analyzing additional dimensions of risk may seem to invite additional debate, but experience shows that extended and unproductive debates have been prompted by omissions in existing analyses. In large and complex decision exercises, risk characterizations that do not consider ecological, social, or human health outcomes that are important to some of the interested and affected parties or that are based on a process that excludes key parties can lead to court challenges and other activities that question the technical adequacy of the analyses, when the actual concern is the process or risks that were never analyzed. We believe this pattern has been a major cause of delay in decisions about high-level radioactive waste disposal, siting waste incinerators, and other intensely controversial risk decisions. A deliberative process that ensures that the decision-relevant risk analyses are performed the first time may reduce delay in such cases, rather than increasing it. If a decision requires additional analyses to meet the major concerns of important parties to the decision, the short-term expense of obtaining the additional expertise may actually be an investment in longer term savings of time and money.

In some cases, there may not even be additional expense. Deliberations in advance of risk analysis may reduce the immediate costs of analysis or increase its cost-efficiency by directing limited resources for analysis to the most decision-relevant issues. For instance, incremental efforts to reduce uncertainty in risk analyses could be directed toward those uncertainties whose reduction might change some of the parties' understandings enough to affect their judgments about what should be done.

We believe that an old carpenter's adage applies to risk analysis and characterization: *measure twice, cut once; measure once, cut twice.* Experience shows that analyses, no matter how thorough, that do not address the decision-relevant questions and use reasonable assumptions can result in long and detailed criticism, great expense and delay in responding to the criticism, and even rejection of a risk decision in court. These hidden costs are especially likely to arise when decisions involve very big stakes and major controversy. When the entire decision-making process is taken into account, not only the costs of analysis, it often costs less to get it right the first time.

It is reasonable to ask whether broadening deliberation, especially on issues that have a strong technical component, such as selecting assumptions for risk analysis, will cause confusion and delay, particularly if the added participants do not understand the technical issues at stake or the language of the technical experts. This potential certainly exists, and avoiding it imposes costs. To involve nonexpert participants meaningfully, efforts must be made to educate them technically or to find individuals who understand the technical issues and can represent the parties' knowledge, perspectives, and concerns in a way that satisfies those parties. The problem of meaningful participation and the costs of achieving it are most serious with parties that have not been well organized or that lack resources to identify or hire their own experts. However, leaving those parties out of meaningful deliberations that affect the risk characterization has its own dangers to the quality of understanding and to the acceptability of the ultimate decisions, as noted in Chapters 2 and 3. These dangers are sufficient in our judgment to warrant experimental efforts to provide resources to allow meaningful participation for parties that could not otherwise join effectively in deliberations. Such experimental efforts should be focused first on risk decisions that may seriously affect the parties in question. They should be designed in consultation with the parties to be assisted and carefully evaluated, with the collaboration of those parties, as to both process and outcome.

Another concern about broadening analysis and deliberation comes from a recurring problem with risk decisions in the United States. Some parties use repeated requests for broader analysis or further deliberation as a tactic to delay a decision or to advance their interests as they could not in the decision process as originally organized. In some cases, a risk characterization process that encourages broader analysis and deliberation will invite such tactics. Yet sometimes those requests for delay are tactical reactions to procedures that excluded some of the parties' chief concerns from consideration on an a priori basis. Delays of this type would become less frequent under a broader concept of risk characterization. Clearly, procedural safeguards are needed to reduce inappropriate or avoidable delays. We encourage organizations to experiment with different approaches to dealing with the problem of strategic delay. We note, however, that in implementing the broader concept of risk characterization, especially in decisions with wide impact, more of the effort of risk analysis and characterization will be focused on substantive questions about risk and less on the adequacy of procedures. Furthermore, broadly based deliberation offers a more efficient forum than the courts for arriving at socially acceptable judgments about whether a request to extend analysis or debate is necessary to the overall decision process.

We emphasize that our proposed approach adds to analysis, deliberation, and participation only as appropriate to specific situations. We do not propose wholesale analysis of every possible outcome, deliberation of every analytical issue, or involvement of all interested and affected parties in all steps leading to a risk characterization. Rather, we advocate that these possibilities be uniformly considered and that as many additional activities included as is appropriate for the situation. For some of the most contentious risk decisions, our approach indeed calls for very extensive analysis, deliberation, and participation—much more than is envisioned in a linear approach that would only allow deliberation after a risk analysis is supposedly complete. However, for most such decisions—such as about high-level radioactive waste disposal, dioxin, and the like—the linear approach has not produced the efficient process that was hoped for, even after huge investments in risk analysis and characterization. For the vast majority of risk decisions, our approach calls for much less than the most extensive possible analytic-deliberative process, and may not add much, if anything, to the current level of effort. We note, though, that the design of simple, generic risk characterization procedures should presumptively involve the spectrum of interested and affected parties and that established processes of this type should be periodically reviewed.

A final concern about the practicality of our approach is that forces external to the analytic-deliberative process may sometimes preclude its use. There may be organized political opposition to allowing some parties a voice in risk analysis or to considering certain kinds of concerns in a risk characterization. This sort of concern is highlighted by the California Comparative Risk Project and the regulatory negotiation on disinfectant by-products (described in Appendix A). In both cases, the conclusions of an analytic-deliberative process with many of the features of our approach was overturned in the larger political system. In California, the argument was that considering certain kinds of risks to human welfare in risk ranking would be unscientific. It is normal in a democracy for parties dissatisfied with a decision to seek redress elsewhere in the system. However, when extensive efforts are made to involve the full range of interested and affected parties in a deliberative process and individual interests bypass the process, it is destructive of the search for deliberative solutions. It may be necessary or valuable at times for the larger political system to legitimate analytic-deliberative processes and thus make it more costly for interests to bypass them. We have not considered the advantages and disadvantages of different ways to legitimate these processes. Among those that might be considered are declaring in advance that the results of certain analytic-deliberative processes will be legally binding on government agencies and establishing analytic-deliberative processes or forums that would continue to participate in governance functions, advising on the final decision, monitoring its implementation and its effects on the things at risk, and recommending changes in policy as appropriate. The latter approach has sometimes been used with local risk decisions about such matters as landfills.

DIAGNOSIS: MATCHING THE PROCESS TO THE DECISION

A key to implementing our approach is to match the analytic-deliberative process to the needs of the risk decision. Doing this can be difficult, and not enough is known to justify any standard procedure for matching. In this chapter, we offer some guidance on how to make that match.

We find it useful to rely on an idea most often associated with medicine, that of diagnosis. A government agency or other organization responsible for risk characterization begins with a diagnosis of the potential hazard situation that is sometimes explicit (e.g., risk defined by law) and sometimes implicit. Diagnosis includes, at minimum, ideas about the nature of the hazard and the hazard situation, the purposes for which the risk characterization will be used, the kinds of information that will probably be needed, and the kind of decision to be made. Diagnosis is typically implicit when an agency applies an existing routine, presuming it adequate for the situation at hand. At other times, important elements of the diagnosis are implicit, such as assumptions about which parties are affected or which threats of harm deserve analysis. We recommend that diagnosis be conducted explicitly far more often than is the current practice. Although a single organization may have responsibility for risk characterization, diagnosis generally benefits from interactions of its staff with scientists, policy makers, and interested and affected parties.

Diagnosis results in a provisional procedure for each step of the analytic-deliberative process leading to a risk characterization. These provisional choices should be reconsidered, with input from the interested and affected parties, during the entire process. Because the best approach to a risk characterization depends on the specific situation, the first element of diagnosis is surveying what we call the risk decision landscape.

The Risk Decision Landscape

Risk decisions vary along many dimensions. Although it might be desirable to reduce these to a few, as has been done with the qualitative aspects of hazards (see Figure 2-5, p. 63), analogous systematic research has not been done on risk decision situations. In our judgment, there is no simple set of categories that can be confidently used to reduce the great variety of risk decisions to a few types for the purpose of designing a few standard approaches to informing the decisions. A similar conclusion has been reached by Graham et al. (1988). We believe, however, that it is useful to consider a number of diagnostic questions before embarking on the processes that lead to risk characterizations. We list some key questions below.

A good diagnosis can help practitioners narrow the range of appropriate courses of action and thus increase the chances of a rational and socially acceptable outcome. For example, past experiences with decision situations similar to the one at hand can provide some guidance on procedures that may work well in the new situation. A good diagnosis can make it easier to consider how a risk decision problem differs from those made in the past and whether existing decision routines should be changed.

Despite our caution about the feasibility of classifying risk decision situations for the purpose of creating a small number of decision-making routines, we think it is useful to keep in mind the following five categories that occur with some frequency on the risk decision landscape. The first category, *unique and wide-impact decisions*, encompasses those that are most well known and most often controversial because they are one-time actions that affect a large portion of the country, a large number of people, or have effects for a very long time. At the other end of the spectrum is the category of *routine and narrow-impact decisions*, which are usually very similar to previous decisions and involve a small geographic area, few people, or primarily short-term effects. Somewhere in the middle of this spectrum is a category that we term *repeated, wide-impact decisions*: like unique and wide-impact decisions, they may have major effects over a large area or large numbers of people, but like routine and narrow-impact decisions they are very similar to previous decisions so that it is relatively easy to anticipate the issues they will raise. Our last two categories do not directly involve decisions about a specific potential hazard: *generic hazard characterizations* and *decisions about policies for risk analysis*. Because these two categories of activities are not designed to result in a specific decision about a specific hazard, they may appear to be outside our framework for an analytic-deliberative process; however, we believe that they, too, can often benefit from this process, beginning with diagnosis.

Unique and Wide-Impact Decisions

Unique and wide-impact decisions are one-time decisions of national or even wider import that usually involve many kinds of interested and affected parties and disparate perspectives on what is at risk. The paradigmatic case is that of decisions associated with the Yucca Mountain site for a permanent national repository for high-level radioactive waste. Because of the size of the stakes in the ultimate decision in such an instance, risk characterization often needs to be based on extensive analysis and deliberation with broad participation or representation of the spectrum of interested and affected parties at every step of the process. Such decisions present special challenges in planning and in carrying out an effective analytic-deliberative process. The process considers risks on a large social and geographic scale, and it often extends over considerable time. It may expand in ways (other topics, across distance) that were not foreseen at the start. Because of the importance of the decisions, considerable resources and time may be available to meet these challenges, so it may be easier to meet the needs for breadth, inclusion, and attention to process than when less weighty issues are at stake.

Routine and Narrow-Impact Decisions

In contrast, some risk characterization procedures support thousands of routine and narrow-impact decisions each year. These may include decisions to issue permits to release small amounts of effluents into air and water, to approve building designs as adequately earthquake resistant, to accept individuals as blood donors, and so forth. Although there may be significant unresolved scientific issues underlying individual decisions, an extended analytic-deliberative process for informing each one would be impractical and might not always serve the overall public interest. For this reason, it is reasonable to routinize the associated risk characterizations.

When establishing routines, it may be useful to consider using a broad-based analytic-deliberative process to devise a general procedure that would be used for the individual decisions and then provide for an appeals procedure for individual decisions as well as periodic public review of the general process. The periodic review should probably involve an analytic-deliberative process roughly as broadly based as the initial effort to create the routine. Periodic review may be useful for a variety of reasons. Such review may be legally mandated; there may be a change in management followed by calls to reconsider certain decisions; occasionally, scientific breakthroughs may indicate a better scientific basis for future decisions; or there may be public calls for change resulting from a process failure. An organization may also desire to change its procedures because of concerns about representation in the process, inefficiencies, or shrinking resources. Because of the potential for loss of trust, the responsible organizations should consider planning for regular review of the risk characterization routines used for informing major classes of routine, quick decisions.

Many of the considerations that apply to setting up and conducting the analytic-deliberative process leading to risk characterizations for unique, wide-impact decisions apply also to reviews of routines for risk analysis or characterization. There are a series of obvious questions to ask in evaluating routine decision-making processes:

• *Input and access*: Do some parties have considerable input to decisions while others have little or none? Are there sufficient resources for adequate participation? Is the process closed because of claims regarding confidentiality? If so, are there mechanisms for review?

• *Decision quality*: Are complaints regarding decisions justified? Are there criteria for identifying bad decisions in the absence of complaints?

• *Efficiency*: Are there mechanisms for staff to identify and easily point out inefficiencies in the process? Are some decisions quickly made that should require more review and deliberation? Are some decisions too slow?

• *Trust and satisfaction*: Do some of the interested and affected parties express distrust about the process? Do some of the parties claim that the decisions are arbitrary or unfair?

• *Review*: When was the last time the process was evaluated or reviewed? Are there established processes for evaluation or review? Is there an appeals process? Is the review process broadly accepted?

• *Resources*: Is the process too costly? Does it have sufficient resources? If not, are additional resources available?

Repeated, Wide-Impact Decisions

Repeated, wide-impact decisions are those that will be the subject of widespread attention because of their possible effects but are sufficiently similar to other decisions that some routinization seems possible. Typical of these are decisions about approving siting and operating permits for power plants and hazardous waste facilities and, in recent years, about strategies for restoring ecosystems. The analyses for them are often local or regional in scope and so may present fewer logistical limitations than national cases since the interested and affected parties may have easier access to the responsible organizations.

These decisions have aspects in common with the previous two categories of decision. They often present the potential to routinize aspects of the analytic-deliberative process, yet they have wide impacts and sometimes high stakes. The responsible organizations should be alert to the need, especially when the likely impact of a decision or the potential for controversy is great, to design or modify aspects of the process to suit unique needs of the particular decision. As with routine decisions, any standard procedures should be periodically reviewed; as with unique, wide-impact decisions, it is important to consider instituting broadly based deliberative mechanisms in one or more of the tasks leading to a risk characterization.

Generic Hazard and Dose-Response Characterizations

Generic hazard and dose-response characterizations are not designed to inform any particular decision, but to serve as inputs for a class of decisions. Examples include efforts to describe the health risks of dioxin, the impacts of climate change on human and ecological health and the economy, and the likelihood of airborne transmission of a particular disease. These efforts are abstracted from the context of any particular decision about a specific situation in which exposures take place, but they can have far-reaching impacts and therefore deserve careful attention, similar to that accorded risk characterizations associated with unique decisions. A special problem with these risk characterizations is that absent a particular decision, it may be more difficult to identify the interested and affected parties in advance and hence to arrive at the most appropriate formulation of the problem. Typically, people in the institutions that would clearly be affected (e.g., producers of a chemical) and those concerned about the precedents that may be set or changed by these types of decisions can be relied upon to participate, but representatives of the more general public or environmental or community groups cannot, perhaps because of lack of resources or limited expertise. Ways to broaden participation in these exercises should be explored. Decisions about setting standards, for instance, for exposure to a toxin or for the performance of a piece of equipment, raise many of the same issues as generic hazard characterizations (Fischhoff, 1984).

Decisions about Policies for Risk Analysis

Decisions about policies for risk analysis are procedural or methodological in nature, such as decisions about which dose-response model to use in toxicological analysis, whether to routinely consider psychological impacts in risk analyses, how to gather information about previously unstudied outcomes, and so forth. These have already received considerable attention from an analytic standpoint from the National Research Council (1994a). These decisions are like generic hazard characterizations in that they are not tied to a specific risk situations but may affect many substantive decisions. Consequently, it may be difficult for many of the interested and affected parties to recognize their importance and to mobilize resources to participate in the analytic-deliberative process. The responsible organizations may need to make special efforts to identify and involve the parties and to ensure broad and balanced participation.

Diagnostic Steps and Questions

To diagnose a risk decision situation, we offer eight steps. Depending on the risk decision, the effort involved may be very brief or rather extensive. This section is written primarily with government agencies in mind, but we believe it will also be useful to other organizations responsible for analyzing and characterizing risks. We do not intend to create a new bureaucratic procedure, but instead to reduce wasted effort through advance thought and planning. Figure 6-1 shows the diagnostic steps involved in preparing to conduct an analytic-deliberative process to inform a risk decision. It represents the fact that the steps are not necessarily sequential—all of them flow into the diagnosis and thus inform preliminary choices about how to execute the process.

1. Diagnose the Kind of Risk and the State of Knowledge

Every analytic-deliberative process sets boundaries as it begins to consider the risk problem, define options, and examine consequences. Considering these boundaries explicitly and systematically from the start has the advantage of identifying mismatches between the boundaries that the responsible organization is initially inclined to set and the demands of the situation.

This phase of diagnosis begins with asking basic questions about the hazard; see box on page 144). Answers to the questions about who or what is exposed have implications for who should participate in the analytic-deliberative process, including the possible need to find ways to include the perspectives of parties that cannot speak for themselves (e.g.,

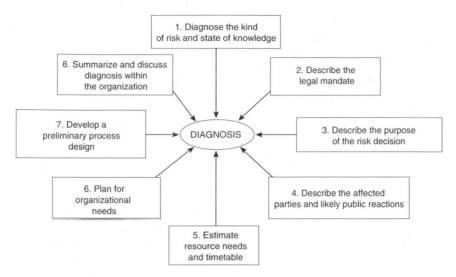

FIGURE 6-1   Diagnostic steps for risk decision making.

infants, future generations) or that lack sufficient expertise to be effective participants. These answers may reveal highly exposed or susceptible populations or suggest that those at greatest risk are not identifiable. If risks appear on initial consideration to be inequitably distributed as a function of race, gender, socioeconomic status, or other factors, the diagnostic effort should lead to a conclusion that the analytic-deliberative process specifically address these issues and that the potentially affected parties participate to ensure that the process is carried out in a way they find competent and credible.

Answers to diagnostic questions about the nature of the harm should reveal the kinds of human health effects and ecological impacts that may need to be characterized and the various other kinds of possible adverse outcomes that technical experts or interested and affected parties consider important. The test for which harms to consider in the analytic-deliberative process is a practical one: Which harms must a characterization address for it to be accepted as sufficiently thorough?

Answers to questions about the qualitative features of the hazard should help in anticipating demands from potentially affected parties for detailed analysis and information, as well as for opportunities to participate. Hazards that are high on dimensions associated with dread and lack of knowledge (the upper-right quadrant in Figure 2-5) are especially likely to generate such demands and to have ripple effects, partly because one such event may signal or portend future, perhaps catastrophic mis-

Diagnostic Questions About Characteristics of Hazards

Who is exposed? Human beings, nonhuman organisms, ecological systems? In the present, or future? Those who benefit from the hazardous activity, or those who gain little noticeable benefit?

Which groups are exposed? Identifiable sensitive or highly exposed populations? Is risk a function of race, class, gender, or occupation?

What is posing the risk? Engineered processes that may fail? Emissions from industry or agriculture? Dangerous behavior? Contaminated drinking water?

What is the nature of the harm? Ecological disruption? Sudden injury or death? Morbidity? Delayed mortality? Harm to wilderness, scenic beauty, sacred sites, religious values? Loss of money or property? Threats to community or democratic process?

What qualities of the hazard might affect judgments about the risk? Perceived voluntariness of exposure, dread, catastrophic potential, signal value of an event, equitable distribution, familiarity of the risk, immediacy of the effect, degree of scientific knowledge available, parties' judgments about who is responsible for causing the hazardous situation?

Where is the hazard experienced? Locally in unique events? Locally and repetitively? Regionally or globally?

Where and how do hazards overlap? Are those exposed also disproportionately exposed to other hazards? Are there synergisms or offsetting relationships among hazards?

haps (Slovic, 1987). Such hazards thus may warrant extra analytic attention and deliberation.

Diagnosis should also consider questions about the state of knowledge about the risk; see box on the next page. There are a few risk decisions for which actuarial data provide a solid knowledge base (e.g., automobile fatalities), but the state of knowledge is more commonly characterized by a mixture of expert judgment, inference, and uncertainty. In such cases, due consideration should be given to how much incremental knowledge additional analysis might buy and what it will cost in terms of time and money, before making initial estimates of the best way to balance analysis and deliberation for the purposes of understanding specific risk factors and clarifying uncertainties.

Answers to the diagnostic questions about consensus and possible omissions from analysis can suggest what will be needed for a risk characterization to meet the likely demands on it. What kinds of analysis will the interested and affected parties demand? What sorts of quantitative

> ### Diagnostic Questions About Characteristics of Knowledge About Risks
>
> *How adequate is the data base on the risk?* Solid knowledge based on repetitive actuarial experience with the specific risk? Some data that must be supplemented with judgment? Few data, with extensive use of scientific judgment required to make estimates? No data, only speculation?
>
> *How much scientific consensus exists about how to analyze the risk?* Is there agreement about analytical methodology, theoretical basis for the analysis, harms not analyzed? Do opinions on these issues differ in a systematic way, for example, by the affiliations or academic disciplines of the scientists?
>
> *How much scientific consensus is there likely to be about risk estimates?* Is epistemic uncertainty a serious problem?
>
> *How much consensus is there among the affected parties about the nature of the risk?* How compelling are experts' explanations for the parties? How stable or volatile is the social understanding of the risk situation?
>
> *Are there omissions from the analysis that are important for decisions?* Are possible harms, management options, or effects (e.g., synergistic or intermedia) left unassessed? Are interested and affected parties likely to consider these omissions serious?

information will be needed on risk and uncertainty? Will there be a need for qualitative descriptions of the state of scientific knowledge? What kinds of expertise and which scientific perspectives will have to be included for the characterization to achieve balance? The diagnosis should result in preliminary answers to such questions.

While the answers to these diagnostic questions are only preliminary, they can lead to a sounder and more credible analytic-deliberative process that addresses, from the outset, most of the issues likely to be of concern. When the process begins, of course, it may alter the organization's preliminary list of options and issues.

2. Describe the Legal Mandate

The legal obligations and legal environment surrounding the decision activity should be recognized or clarified at the beginning. Issues to consider for a government agency include the agency's legislative mission; legal requirements affecting the decision-making process (e.g., Administrative Procedure Act); the degree to which the decision-making authority has been legally delegated; the demand for evidence in terms of burden-of-proof requirements; expectations regarding legal challenges; requirements for documentation of the process and for tangible analytic outputs; and the legal responsibilities of interested parties both before and after the decision.

Because of the importance of broadly based deliberation, an agency should avoid taking the stance that a statute prohibits public input simply because it gives the agency full decision-making responsibility. Rather, it should develop a clear understanding of how much statutory discretion it can exercise in order to listen to issues as needed without abdicating responsibility. Agencies should avoid giving the impression that they will prolong a process until all participants are satisfied or that listening to the interested and affected parties is equivalent to a commitment to decide in favor of those who testified.

3. Describe the Purpose of the Risk Decision

The responsible organization's staff should describe the stated and implicit purposes of the decision-making activity, the type of decision and general aims furthered by the activity, and the intended users of the risk characterization. Different types of decisions may require different types of knowledge and perspectives and hence require different participants in the analytic-deliberative process—both inside and outside the organization. Consider the variety: Is the decision about risk analysis technique (e.g., selection of default assumptions), about guidelines for making inferences from data, about regulating an industrial process, about setting an emissions standard, about taxing emissions and effluents, about establishing cancer potency values, about informing individuals at risk, about policy strategies, or about implementation? Different kinds of decision also affect different parties, whose concerns the process must satisfy and whose participation or representation it may require.

Staff should identify the types of decisions that will probably follow the risk characterization and consider how the particular risk characterization activity will facilitate the decision and the overall aims of the organization. It is important to also consider possible secondary and tertiary impacts of the decision, which may be of more concern to interested and affected parties than the primary one.

4. Describe the Affected Parties and Likely Public Reactions

Diagnosis should consider the identity and likely positions and perspectives of the interested and affected parties. The first step is to provide a tentative identification of the parties and any barriers there might be to their effective involvement in the analytic-deliberative process. As already noted, the success of the process depends on the satisfaction and appropriate participation of these parties. It may be difficult to achieve meaningful participation of some of the parties in those parts of the process involving discussion of complex technical issues because of lack of expertise and funding of those parties, because they mistrust the responsible organization and are unwilling to participate, or because they trust the organization implicitly and see no need to participate. Government agencies should resist the temptation to exclude parties whose views are known to be different from those of agency administrators.

Diagnosis should consider the possibility that some affected parties may resist participation because they believe they are more likely to achieve their desired outcomes by some other strategy, such as a legal challenge. Diagnosis should result in tentative recommendations on how to address any such problems that seem likely to become significant. (We discuss some of the possibilities in Chapter 3.)

Diagnosis should also consider the potential for controversy. Which parties are likely to support or oppose a possible decision? How might they try to exert influence indirectly? How strong is the public consensus on the need to address the hazard? What type of press coverage can be expected? It is important to assess the potential for "rough weather," consider the organization's possible responses to it, and lay the groundwork for a strategy for addressing external pressures. Informal contact with interested and affected parties can provide valuable insight on these issues and on how to address them. In some cases, a reorientation of the analytic-deliberative process can help reduce controversy or channel it more productively.

The indicators of strong potential for public controversy include:

- There is a clear distinction between those people who benefit from one of the decision options and those who do not. The potential is strongest when the distinction reinforces preexisting social divisions (e.g., rich versus poor, workers versus employers, regional or racial differences).
- Recent decisions that are similar to this one evoked controversy.
- The responsible organization suffers from a low level of public trust or lacks a public constituency.
- The hazard is characterized by a high degree of dread or the potential for widespread, involuntary exposure.
- The issue presents news "hooks" that attract media attention or vivid opportunities for use in larger debates (e.g., incinerator siting and the debate on pollution prevention strategies; the presence of endangered species).

There are numerous other factors that may also be worth considering (Foran et al., 1995). The same factors also suggest the probable nature of the controversy that the organization may face.

In our view, the effectiveness of a risk characterization will depend significantly on the prevailing climate of public opinion. It is therefore imperative for public officials to diagnose that climate so that they can arrange for appropriate participation and direct scientists to address issues likely to be raised as criticisms if they are omitted from the risk characterization.

Conflicts over substance and process are closely related. Affected parties that raise objections to substantive conclusions or omissions in a risk characterization product are often also reacting to perceived failings in the process. If people do not trust the process, they have little trust in the outcome. Therefore, when public controversy can be anticipated, diagnosis should pay particular attention to the concerns of the affected parties and to including those parties in the process.

We emphasize again that understanding the potential for controversy and designing an analytic-deliberative process accordingly are not enough to prevent some unwanted outcomes. Deliberations across the range of decision participants may fail to reach consensus and, sometimes, interested or affected parties may choose to exert influence indirectly and outside the officially designated process, such as through litigation, legislation, or mass media publicity. Even with good diagnosis and planning, such reactions may occur, but they are likely to be less incapacitating.

5. Estimate Resource Needs and Timetable

Diagnosis should consider the internal and external time pressures on the decision and the extent to which they are explicit. It should consider the public health and other consequences of indecision, and if significant, the risk-reduction potential of interim actions during a lengthy decision-making process (Harris, 1990). It should consider whether additional resources might be made available or whether resources might be cut. The adequacy of resources depends in part on the range of expertise needed for analysis and on the potential for controversy. Evaluating the timeline in terms of the legislative, budgetary, and executive election cycles is a must for complicated decisions by government agencies.

6. Plan for Organizational Needs

The diagnosis should result in a plan that specifies the organizational support needed for the risk characterization. When a government agency is responsible, it should consider the needs for coordination between the program unit given initial responsibility and other units within the agency, other agencies, and other levels of government. It should explicitly consider the need for early coordination with units that maintain regular contact with interested and affected parties (e.g., risk communication programs). Several diagnostic steps can benefit from the input of in-house experts, as well as the informal input from interested and affected parties themselves. The diagnostic process should consider the need for a task force or some similar entity, focused on the particular risk decision, that cuts across the usual organizational structure. If this sort of coordination is needed, it should be implemented early.

7. Develop a Preliminary Process Design

The diagnosis should result in a clear proposal for the steps of the analytic-deliberative process, their sequence, expected iterations, participants, rules for closure and other decisions, and tangible products. The plan should be open for discussion by the affected and interested parties once the process begins, and it should be changeable as needed. It should consider the legal and resource constraints on the process, where and how affected and interested parties can participate, time commitments, and overall time frame. It should also clearly specify whether tangible products will be needed describing the risk and documenting the process.

8. Summarize and Discuss Diagnosis within the Organization

Both "risk managers" and "risk assessors" should be actively engaged in all parts of the diagnosis. The discussion and review of the many judgments involved in the diagnosis will help to surface potential problems within the organization, clarify the degree of commitment the organization should make to the activity, and ensure that the organization enters the process with a consistent position on what it is willing to do in terms of participation, deliberation, and other potentially contentious issues.

Conclusion

Diagnosis should result in a commitment within the responsible organization, among both staff and management, regarding the nature and level of effort of the analytic-deliberative process leading to a risk characterization. Especially when the risk analysis is expected to be complex and difficult and there is likely to be polarization and politicization about the risk decision, an understanding among both staff in the organization and the interested and affected parties that the organization is committed to the activity is essential. The diagnostic effort should therefore result in explicit expectations about the extent of the activity and the kinds of support and constraints that will come from the organization.

The responsible officials should treat the diagnosis as tentative. One of the greatest dangers of diagnosis is that it may convey a sense that the problem formulation, the process design, and other aspects of the analytic-deliberative process are firmly established. Diagnostic efforts are subject to the test of experience. An unwillingness to modify preliminary decisions can undermine the larger purpose of making risk characterization responsive to the emerging needs of the decision makers and the interested and affected parties. Officials of the responsible organization should always keep in mind that their goal is a process that leads to a useful and credible risk characterization.

BUILDING ORGANIZATIONAL CAPABILITY

Implementing effective risk characterization requires appropriately structured and staffed organizations and systematic efforts to improve the knowledge base for designing and managing the analytic-deliberative process that informs risk decisions.

Organizational Issues

Implementing a broadly based analytic-deliberative process for risk characterization makes demands on an organization. It must assign responsibility and authority for diagnosing risk decision situations and for implementing new analytic-deliberative processes, and it must create procedures for reviewing and approving its process decisions. The relevant staff must understand the underlying concepts, which may require special training. The organization must be prepared to respond appropriately to requests to analyze dimensions of risk it has not analyzed before and to acquire the necessary expertise to do so. It must also be prepared to cope with the possibility of attempts by some of the interested and affected parties to delay a decision, and it must develop a range of strategies for reaching closure on decisions that affect the process leading to risk characterization.

Broadening the process also requires new kinds of coordination between the organizational units normally responsible for risk analysis and those responsible for interactions with interested and affected parties. This coordination may require changes in organizational structure or procedures. It may be advisable in some situations to establish task forces or working groups that cut across organizational units so as to involve all the relevant units in the analytic-deliberative process, beginning with the initial decision to characterize a risk, and to maintain communication throughout.

For some organizations, a broad concept of risk characterization may require changes of procedure to permit flexibility and judgment. Those responsible for managing the process that leads to a risk characterization must be allowed the flexibility necessary to match the process to the decision situation. Their organizations will need to develop ways to allow that flexibility and at the same time guard against arbitrary decisions and undue influence from interested parties. Replacing a reliance on standard procedures with a more flexible system will require care both in assigning responsibility and in establishing safeguards. Perhaps most important, organizations should develop mechanisms that provide feedback on their procedures so that they can be improved over time. Some agencies consider reviewing deliberative innovations to be integral to their success and improvement (Fisher, Pavlova, and Covello, 1991; Young, Williams, and Goldberg, 1993; Grumbly, 1996). (On organizational learning related to risk decision making, see Short and Clarke, 1992; Chess, Tamuz, and Greenberg, 1995.)

Improving the Knowledge Base

Only a very limited knowledge base exists for guiding decisions about the process that leads to risk characterization. Thus, organizations that modify their standard procedures and adopt a carefully designed analytic-deliberative process must to a great extent find their own paths. This situation can be improved over time if explicit and systematic efforts are made to evaluate and learn from experience. Such efforts can help organizations conserve resources and solve problems in at least three ways:

- Gathering knowledge and feedback early and throughout the analytic-deliberative process allows mid-course corrections that save time and money.
- Pretesting materials that summarize risk information reveals in advance whether this information is understandable and useful for the intended audiences.
- Retrospective analysis can suggest ways to improve future efforts (Office of Cancer Communications, 1989; Fisher, Pavlova, and Covello, 1991).

Ideally, an organization involved in a major analytic-deliberative process will devise systems of feedback and evaluation to inform it both during and after the process. In addition, institutions that provide scientific support for many of these organizations, such as federal scientific agencies and industry-based research institutes, should support systematic efforts that build knowledge about analytic-deliberative processes that may have general value for many organizations.

One of the difficulties in building knowledge about analytic-deliberative process is defining criteria for success. We do not suggest that interested and affected parties or organizational managers must all be "happy" at each step of the process for it to be considered successful. A more realistic approach is to define criteria for success early on, at or before the step of process design, using a process in which both interested and affected parties and organizational staff participate (Rosener, 1981). We believe that asking questions like those listed below will yield valuable insights that can be used to develop realistic expectations for the analytic-deliberative process and to arrive at a working definition of success or effectiveness.

Criterion	Measurement procedure
Getting the science right	Ask risk analytic experts who represent the spectrum of interested and affected parties to judge the technical adequacy of the risk-analytic effort
Getting the right science	Ask representatives of the interested and affected parties how well their concerns were addressed by the scientific work that informed the decision
Getting the right participation	Ask public officials and representatives of the interested and affected parties if there were other parties that should have been involved
Getting the participation right	Ask representatives of the parties whether they were adequately consulted during the process; if there were specific points when they could have contributed but did not have the opportunity
Developing accurate, balanced, and informative synthesis	Ask representatives of the parties how well they understand the bases for the decision; whether they perceived any bias in information coming from the responsible organization

Evaluation or feedback should take a form appropriate to the scale and nature of the analytic-deliberative process: a resource-intensive risk characterization will merit more rigorous and extensive evaluation than a more limited one. Evaluation efforts may use quantitative or qualitative methods and aspire to different degrees of rigor as the situation demands. (For detailed discussion of principles and methods of evaluation for decisions involving broadly based participation, see, e.g., Sewell and Phillips, 1979; Rosener, 1981; Fisher, Pavlova, and Covello, 1991; Syme and Sadler, 1994.) Generally, conscious efforts at evaluation both during and after the process are important for improving analytic-deliberative processes.

Feedback and evaluation can begin in the diagnosis phase, when the responsible organization begins to define the resources it will need to develop a risk characterization and to develop a preliminary process design. Process evaluations that seek data for midcourse corrections should be common practice. The U.S. Department of Energy, for example, has developed a practice in which citizens' advisory committees do self-assessments on at least a yearly basis (Grumbly, 1996). Surveys have also been used to solicit feedback before and during efforts to involve communities in problem solving (e.g., Pflugh, no date). Informational materials can be pretested for comprehensability and relevance and to ensure that they deliver the intended message, as well as being technically accurate (e.g., Office of Cancer Communications, 1989; Morgan et al., 1992). We strongly suggest that risk characterization messages developed for wide distribution undergo pretesting with the intended audience.

Other innovative approaches may also be appropriate in certain situations. One is to conduct simulations to illuminate particular issues. Simulations mimic actual decision processes but are conducted outside the decision itself. They can be used to examine the ways that deliberative groups formulate problems, identify the outcomes that require analysis, and interpret scientific information. Simulations can be useful for identifying potential problems with existing processes and for suggesting particular approaches that might be tried in real decision contexts. (For an illustrative example, see Hester et al., 1990). Organizations may also use quasi-experimental evaluation designs to compare different procedures, for instance, for integrating analysis into deliberations, for discussing scientific information in a diverse deliberative group, for arriving at particular judgments to be incorporated as assumptions in risk analyses, or for reaching closure. In this approach, an organization treats its innovations as experiments and studies their effects in comparison with other procedures. An organization might also use what it believes will be an exemplary process and gather data on it. If the first attempt is successful, it might become a benchmark for future processes; regardless of its degree of success, it can become a baseline for learning and improvement.

An organization can use informal feedback or combine formal and informal evaluations. For example, it might organize broadly based advisory groups to review the processes it has used to make the judgments at each step leading to risk characterizations. The advisory groups could consider whether it appears with hindsight and with the findings of evaluation research that the analytic-deliberative process might have been more effective if conducted differently. An organization, with the help of its advisory group, might then consider what lessons past experience holds for future risk characterization efforts. Organizations might also build and update libraries of case files that use standard categories to record the processes used to make the judgments that affect its risk characterizations. These files would provide a basis for systematic case study research on the organization's past experience and for review by an advisory group.

CONCLUSION

A broadly based analytic-deliberative process may require organizations that characterize risks to engage in new and unfamiliar activities. Although there are legitimate concerns about the practicality of making such changes, we see good reason to do so, while testing the effects on the overall process and implementing safeguards against attempts at tactical delays. Although there will be initial increases in the cost in money and time for some risk analyses and characterizations, the overall costs of risk decision making may actually decrease. It may cost less in the long run to do it right the first time.

Successful implementation depends on matching the analytic-deliberative process to the needs of the decision. This requires a clear understanding of the decision milieu. Although there is no standard procedure for doing this, organizations can benefit by asking a series of diagnostic questions when they plan the process and by keeping their diagnoses flexible and responsive to information that emerges during the process. Implementation may also require organizational efforts at staffing and training and organizational changes to permit the necessary coordination among units and to allow flexibility in the processes informing risk decisions. There is also a need for evaluation research to improve the analytical-deliberative processes on which risk characterization depends.

7

Principles for Risk Characterization

Government agencies and other organizations are increasingly making decisions in which they explicitly consider risks of harm—to human health, safety, and well-being and to nonhuman organisms and ecological systems—along with the other considerations that enter into those decisions. In so doing, they have relied increasingly on analytic techniques developed to establish a solid factual basis of understanding of those risks. Such risk analyses sometimes yield information of a type or in a form not directly useful to decision makers. The term risk characterization is commonly used to describe efforts to make the state of knowledge relevant to a risk decision intelligible to decision participants who may or may not be expert in the techniques of risk analysis. We undertook this study to advise federal government agencies and others on ways to improve those efforts.

Our study has led to a conception of risk characterization as the product of a decision-driven, analytic-deliberative process and to a set of principles for organizing the process. The purpose of risk characterization is to improve the understanding of risk among public officials and interested and affected parties in a way that leads to better and more widely accepted risk decisions.

1. Risk characterization should be a *decision-driven activity*, directed toward informing choices and solving problems.

Scientific efforts in support of risk analysis have sometimes been criticized for being of little help for decision making, even when they have added to scientific knowledge. Effective risk characterization must accurately translate the best available information about a risk into language nonspecialists can understand. But it must also do more. It must address the right questions—the ones that the various participants in risk decisions want answered as a basis for making choices—and it must give those parties an understanding of the many facets of risk. Good risk characterization results from a process that not only gets the science right—that is, involves an adequate level of scientific inquiry and analysis—but also gets the right science—that is, directs that analysis to the most decision-relevant questions.

- *Risk characterization serves the needs not only of the designated decision makers but also of the spectrum of parties that participate in risk decisions.* Although risk characterizations are often completed for the benefit only of an organization's decision maker, it is important to recognize that various other parties have a right to participate in the decision and may do so, either before or after the organization acts. These parties include legislators, judges, industry groups, environmentalists, citizens' groups, and a variety of others. Acceptance of risk decisions by the interested and affected parties is usually critical to their implementation. Satisfactory risk characterization processes and products provide all the decision participants with the information they need to make informed choices, in the form in which they need it. A risk characterization that fails to address their questions is likely to be criticized as irrelevant or incompetent, regardless of how carefully it addresses the questions it selects for attention.
- *Risk characterization should not be an activity added at the end of risk analysis; rather, its demands should largely determine the scope and nature of risk analysis.* It is well recognized that risk characterization depends on good scientific analysis. It is not so well appreciated that risk analysis depends on risk characterization: that the need for characterization to be decision relevant determines which analyses are worth doing. Risk characterization requires a solid scientific base, but it will fail if it does not incorporate the knowledge and perspectives of the various participants in decisions and seriously address the issues they see as critical. Consequently, risk analysis should not proceed very far as the task only of analysts. The scientific agenda should also be guided by and periodically recalibrated on the basis of input from the interested and affected parties.

2. Coping with a risk situation requires a *broad understanding* of the relevant losses, harms, or consequences to the interested and affected parties.

Risk analyses are often restricted to the examination of a narrow set of outcome conditions: certain human health hazards and sometimes a narrow range of effects on ecosystems or on economic interests. Such a focus is sometimes defended on the ground that the outcomes examined are the most serious ones that meet some a priori definition of "risk." This is not an appropriate justification because risk characterization is a goal-directed activity the success of which depends on the satisfaction of decision participants. Relevance to a decision, and therefore to a risk characterization, cannot be determined a priori by a formal definition of risk. The outcomes that should be considered relevant depend on the decision.

- *Risk characterizations should, when appropriate, address social, economic, ecological, and ethical outcomes as well as consequences for human health and safety.* Human health and safety are often the only important consequences to the interested and affected parties, but for some decisions, other outcomes are as important or even more so. Some of the interested and affected parties will feel inadequately informed unless these other outcomes are addressed. A wide range of outcomes is amenable to systematic analysis, although many of them require additional expertise to that needed for assessing outcomes for human health and safety. Even when a decision-relevant outcome cannot be analyzed in a systematic and replicable way, it should still be addressed in the risk characterization so as to avoid leaving the impression that the outcome has been judged to have zero risk.
- *Risk characterizations should, when appropriate, address outcomes for particular populations in addition to risks to whole populations, maximally exposed individuals, or other standard affected groups.* Depending on the decision at hand, adequately informed choice may require that risks be characterized for certain specially exposed or vulnerable populations, such as children, members of particular occupational groups, residents of highly exposed areas, those with increased susceptibility due to genetic or environmental factors, people with compromised health, and groups defined by race, ethnicity, or income.
- *Adequate risk characterization depends on incorporating the perspectives and knowledge of the spectrum of interested and affected parties from the earliest phases of the effort to understand the risks.* If a risk characterization is to illuminate the relevant facets of a risk decision and be credible to the interested and affected parties, it must address what these parties believe may be at risk in the particular situation, and it must incorporate their specialized knowledge. Often, the best way to do this is by the active involvement or representation of the parties.
- *The breadth of analysis and the appropriate extent of involvement or*

representation required for satisfactory risk characterization is situation depen-dent. Many risk characterizations can be satisfactorily completed by ana-lyzing only a few outcomes and with little direct involvement of inter-ested and affected parties, but others require a more inclusive and more participatory approach. The level of effort required is situation specific. What is always required is that the necessary breadth of analysis, charac-terization, and involvement be considered explicitly for each step of the process and not made by default.

> **3. Risk characterization is the outcome of an *analytic-delibera-tive process.* Its success depends critically on systematic analy-sis that is appropriate to the problem, responds to the needs of the interested and affected parties, and treats uncertainties of importance to the decision problem in a comprehensible way. Success also depends on deliberations that formulate the deci-sion problem, guide analysis to improve the decision partici-pants' understanding, seek the meaning of analytic findings and uncertainties, and improve the ability of interested and affected parties to participate effectively in the risk decision process. The analytic-deliberative process must have an appro-priately diverse participation or representation of the spectrum of interested and affected parties, of decision makers, and of specialists in risk analysis, at each step.**

Risk characterization requires a sound scientific base, supported by systematic analysis. Of critical importance is maintaining the integrity of the analytic process; in particular, protecting it from political and other pressures that may attempt to influence findings or their characterization so as to bias outcomes.

Analysis, like all of risk characterization, should be decision driven and aimed at a comprehensive understanding of relevant factors. Analy-sis includes not only the use of systematic methods from the physical, mathematical, and health sciences, but also, whenever relevant for under-standing, analytic methods from the social sciences, ethics, and law. The best available analytic methods should be used, whether quantitative or qualitative, within limits of effort determined by the degree of detail or precision appropriate for the decision. Analysis should address the ques-tions that decision participants need to consider in order to make informed choices, which may require conducting analyses beyond the ordinary scope of risk analysis. Simple and narrowly focused analytic procedures may be appropriate, however, for large numbers of routine decisions if such procedures are justified by prior analysis and delibera-tion and are subject to appropriate review (see Chapter 6).

Uncertainty should receive clear and comprehensible treatment in risk characterization. Participants in risk decisions need to understand both the magnitude and the character of uncertainty: for example, whether it is due to inherent randomness, lack of knowledge, or disagreements of theories, models, values, or perspectives. Useful analytic methods exist for characterizing certain types of uncertainty. They should be focused on the uncertainties that matter most to the decision. Moreover, they should be used with great care. Pervasive and persistent cognitive biases, as well as a variety of social factors, can exert pressure toward misperceiving uncertainty, even when it has been carefully analyzed. The evaluation of uncertainty can enlighten the decision process, identifying those studies and data collection efforts that can most effectively reduce the uncertainties that matter.

Deliberation is as critical to risk characterization as analysis, although its importance has been underappreciated. Deliberation is needed to frame, and where necessary reframe, the decision problem, define the fundamental questions that risk characterization needs to address, set the research agenda, decide who will participate in the effort to build understanding, identify the relevant information, settle on ways to gather the information, select assumptions to use when data are insufficient, and arrive at judgments about the degree of reliance that should be attached to the results of risk analyses and about the amount and kind of uncertainty these results contain. For potentially controversial risk decisions, deliberation should involve the spectrum of interested and affected parties to bring the analysis into better alignment with the parties' needs for information, choose more realistic and satisfactory assumptions based on specialized knowledge the affected parties may uniquely possess, and subject analyses to critical review from a fuller range of perspectives. Broadly based, appropriately participatory deliberation benefits understanding by ensuring that analysis draws on the full range of relevant knowledge and perspectives available in the society, and it benefits the decision process by making it more inclusive and more credible, furthering democratic norms.

Deliberation in the context of risk characterization, even when highly participatory, differs from what is usually called "public participation" in three major ways. First, it precedes agency proposals and action: it is aimed at improving understanding of risk situations, as distinct from taking action on them. Second, because the deliberation is intended to improve understanding, the involvement of knowledgeable experts as well as "the public" is essential throughout. Third, deliberation is not merely a forum in which interested citizens can be heard, but a symposium in which risk experts, public officials, and the various interested and affected parties can interact as equally valid contributors.

Effective deliberation can affect the acceptance of risk characterizations. It can clarify the areas of consensus and disagreement among interested and affected parties about how the problem is framed, how data are interpreted, and what further analysis is needed, thus focusing the characterization on the issues agreed to be most critical to a risk decision. It can promote mutual exchange of information among interested and affected parties and increase mutual understanding. It can reduce problems of mistrust if the responsible organization involves the spectrum of parties and responds to the participants' suggestions about the risk analysis. It can limit conflict by arriving at substantive and procedural agreements, such as about which assumptions to use in analysis or which technical consultants to select. And it may help the participants learn ways to interact productively that they can employ in future analytic-deliberative processes.

- *Organizations should start from the presumption that both analysis and deliberation will be needed at each step leading to a risk characterization.* We are not advocating that explicit deliberation occur at every step of every process. It is sometimes appropriate, for example, to conduct generic deliberations to arrive at analytic procedures that will then be used routinely in a large number of subsequent risk characterizations. However, adequate justification should be given for restricting deliberation. An organization may show, for instance, that sufficient deliberation for the current purpose was done in establishing the appropriateness of an analytic routine being followed.

- *Organizations should start from the presumption that the spectrum of interested and affected parties will be involved in deliberations in each step leading to a risk characterization.* Organizations should consider ways to broaden participation and accommodate demands for it; the burden for justification should be placed on those who would restrict participation rather than on those who would broaden it. Full participation at each step is particularly important when the contemplated risk decision is expected to be controversial or to have widespread and potentially serious effects on many people or areas.

- *Broadly based, appropriately participatory deliberation does not necessarily mean the inclusion of every interested and affected individual and group.* Depending on the purpose of the deliberation, appropriately broad participation may be achievable through the use of surrogates or representatives who bring to the table knowledge, perspectives, and concerns of the parties that are relevant to the issue at hand.

- *Broadly based, appropriately participatory deliberation will sometimes require that resources be provided to some of the interested and affected parties.* Some of the parties to some risk decisions cannot afford the time, the travel, or the technical assistance they need to participate meaningfully in particular deliberations, and it will be necessary to provide such resources in order to obtain the benefits that deliberation can provide for informing risk decisions and increasing their acceptance.

• *An effective analytic-deliberative process depends on explicit attention to process design, especially concerning the deliberation and its integration with analysis.* The deliberative parts of the process require as much advance planning as the analytic ones. Planning should consider who should be involved, the form the process might take, resource needs, timing, and the coordination of deliberation and analysis.

4. Those responsible for a risk characterization should begin by developing a provisional *diagnosis of the decision situation* **so that they can better match the analytic-deliberative process leading to the characterization to the needs of the decision, particularly in terms of level and intensity of effort and representation of parties.**

Risk situations vary along many dimensions, and the same analytic-deliberative process is not appropriate for all risk characterizations. In particular, the level of effort that should go into problem formulation, process design, and the other elements of the analytic-deliberative process—and into securing appropriately broad participation—is situation dependent. Responsible organizations should seek to match the level of effort to the needs of the task. Past experience shows that government agencies and other organizations are more likely to err on the side of inadequate participation and too-narrow deliberation, a bias that should be reversed. Nevertheless, we do not advocate unlimited efforts to broaden the process.

Diagnosis helps organizations use their resources efficiently and effectively. An organization may decide that a decision can be informed quite adequately by following an existing standard procedure for characterizing risks. However, if this diagnosis is incorrect, the organization may generate avoidable ill will and make it difficult to get the needed participation later on. Organizations will do better to err on the side of broadening participation, beginning at the stage of problem formulation, when it may be possible to determine whether a simple procedure for the rest of the process will meet the needs of the parties.

• *For many decisions, a simple, generic risk characterization procedure will suffice.* Many risk decisions do not require extended attention to problem formulation, process design, and so forth within the analytic-deliberative process. With many routine regulatory approvals, for example, it is sufficient to design a single generic risk characterization process that can be used for a large number of specific decisions. However, when a simple or routine procedure is being contemplated, careful consideration should be given to the process design at the outset, the deliberation about process should presumptively involve the spectrum of interested and affected parties, and a commitment should be made to reconsider the procedure from time to time to ensure the adequacy and appropriateness of the routine and to check on the adequacy of representation of the parties. That is, a routine for informing risk decisions should be periodically reviewed through an analytic-deliberative process.

- *An inappropriate or inflexible decision to use a narrow, routinized, or nonparticipatory analytic-deliberative process for risk characterization can undermine the decision-making process.* Some of the most contentious risk controversies have centered on claims of inadequate analytic attention to valid concerns or failure to meaningfully involve some of the interested and affected parties. Examples include controversies over siting hazardous waste facilities and over pesticide spraying in residential areas. Explicit attempts to diagnose the risk decision situation and design the analytic-deliberative process accordingly can make risk characterizations more credible and thereby reduce controversy.

5. The analytic-deliberative process leading to a risk characterization should include early and explicit attention to *problem formulation*; representation of the spectrum of interested and affected parties at this early stage is imperative.

Some of the worst examples of risk decision making have roots in the way that problems were formulated for risk analysts. Difficulties arise predictably, and conflicts are exacerbated, when a large-scale analytical effort is addressed to a problem that some of the interested and affected parties do not recognize as the relevant one for the decision. It is therefore extremely important for the organizations responsible for risk decisions to investigate whether there are or might be competing definitions of the risk problem. Risk characterization can be fairly straightforward if the interested and affected parties agree on which issues deserve analysis; if they do not agree, it is often worth making special efforts at the outset to engage them in deliberation about what should be analyzed.

We do not imply that extensive efforts at participatory problem definition are always warranted. However, failure to make such efforts when they are appropriate can be extremely costly. When there are major differences among the parties in their understandings of the decision problem, it is a serious mistake to proceed without addressing these differences. When problem formulation is given short shrift, there may be a loss of understanding, a loss of credibility for the responsible organization, a serious and possibly avoidable escalation of controversy, delay or paralysis in the decision process, and subsequent economic and social costs of delay to government and society.

6. The analytic-deliberative process should be *mutual and recursive*. Analysis and deliberation are complementary and must be integrated throughout the process leading to risk characterization: deliberation frames analysis, analysis informs deliberation, and the process benefits from feedback between the two.

As already noted, a typical criticism of risk characterizations is that the underlying analysis failed to pay adequate attention to questions of central concern to some of the interested and affected parties. This is not so much a failure of analysis as a failure to integrate it with broadly based deliberation: the analysis was not framed by adequate understanding about what should be analyzed. Risk characterization can also make the opposite sort of error, for example, by addressing decision options that careful analysis would show to be impracticable. This would not be so much a failure of deliberation as a failure to inform deliberation with good analysis. Risk characterization fails when analysis is not properly guided to address the informational needs of participants in risk decisions; it also fails when deliberation is not adequately informed by analysis.

Analysis and deliberation ideally improve each other. Analysis enhances deliberation by informing discussions with facts, predictions, and basic understanding of risk-generating processes. Deliberation enhances analysis in several ways. Deliberation among technical experts can help clarify areas of consensus and dispute and the underlying reasons. Deliberation among public officials, analysts, and interested and affected parties can define needs for analysis and improved understanding. And deliberation can bring new information and new perspectives to analysis. Risk characterization benefits from mutual and recursive interaction between analysis and deliberation and between analytic specialists and the other decision participants.

Both analysis and deliberation have a place in each step leading up to risk characterization: formulating the problem, designing the process, selecting options and outcomes, generating and interpreting information, and synthesizing the state of knowledge. The organizations responsible for risk characterization need to give special attention to the role of deliberation in each of these steps: whose input would advance the task, how that input should be elicited, how it should be informed by analysis, and how it should feed into further analysis.

The interplay between analysis and deliberation sometimes gives reason to revisit past decisions. Deliberation may identify an additional policy option whose effects on the risks need to be analyzed, or an analysis may identify a previously unrecognized aspect of a hazard, so that the meaning of risk information needs to be reinterpreted. In these and other ways, the analytic-deliberative process is recursive, recovering old ground, but with improved understanding.

7. Each organization responsible for making risk decisions should work to *build organizational capability* to conform to the principles of sound risk characterization. At a minimum, it should pay attention to organizational changes and staff training efforts that might be required, to ways of improving practice by learning from experience, and to both costs and benefits in terms of the organization's mission and budget.

Organizations may experience difficulties in following these principles, particularly in regard to increasing input from some interested and affected parties, involving nonscientists in deliberations about risk analysis, broadening the range of adverse outcomes to consider in risk analysis, more fully integrating analysis and deliberation, and doing anything that appears to prolong the decision process or increase its complexity. We are sensitive to concerns about cost and delay, but note in response the massive cost and delay that have sometimes resulted when a risk situation was inadequately diagnosed, a problem misformulated, key parties excluded, or analysis not integrated with deliberation. We believe that following the above principles can reduce delay and cost as much as or more than it increases them.

As a general matter, we believe it is critical for organizations to have the capability to organize the full range of analytic-deliberative processes, including the broadly participatory ones that risk situations sometimes warrant. Organizations differ too much from one another to allow us to make any universal recommendations about how to establish and maintain this capability, but we offer several points for organizations to consider:

• *Having the capability to organize the full range of analytic-deliberative processes may require special efforts to train staff.* Training may be warranted to introduce concepts such as broadly based, appropriately participatory deliberation, integration of analysis and deliberation, and social and ethical risk. It may also be useful for establishing good working relationships between agency units that have not previously collaborated successfully, but that must do so to integrate analysis and deliberation.

• *It may be necessary to acquire analytic expertise with regard to ecological, social, economic, or ethical outcomes.* Experts in analyzing human health risks are not usually expert in analyzing these other outcomes. Additional experts should be involved when these outcomes are important to a risk decision.

• *Having the capability to organize the full range of analytic-deliberative processes may require organizational changes.* For example, it may be advisable in some situations to establish task forces or working groups that cut across units of the organizational structure so as to involve risk analysts, policy makers, risk communication specialists, and others in the analytic-deliberative process, beginning with the initial diagnosis of the problem. It may also be helpful in some agencies to make organizational changes that facilitate creating such cross-cutting groups.

• *Future risk characterizations will benefit from organizations' evaluations of their current activities.* Experience provides a good base for learning how to better diagnose risk decision situations and what works in each type of situation. Organizations involved in analytic-deliberative processes should devise systems of feedback and evaluation to inform them both during and after these processes, and institutions that provide scientific support for many such organizations, such as federal scientific agencies and industry-based research institutes, should support systematic efforts that build knowledge about analytic-deliberative processes and that may have general value for many organizations. Organizations that characterize risk should work with interested and affected parties to define criteria for evaluating the process leading to risk characterization. They should also consider implementing explicit practices to promote systematic learning from their efforts to inform and make risk decisions. These might include establishing broadly participatory panels or advisory groups to review past analytic-deliberative processes, building libraries of case files that use standard protocols for describing past efforts and their outcomes so that experience can provide a basis for learning, conducting formal evaluation research projects to understand and learn from the ways the outcomes of analytic-deliberative processes are affected by how the processes are organized, and using simulations and quasi-experimental research to gain deeper understanding of the analytic-deliberative process.

• *The breadth of focus and participation of analytic-deliberative processes should be considered in terms of the potential benefits and costs to an organization's mission and budget, and to society.* There are obvious potential costs to making risk characterization broader in terms of who participates, which risks are examined, and how extensively deliberation is integrated into the process. These costs are mainly in terms of time and money. The potential benefits, though sometimes less obvious and immediate, may be considerable for controversial or wide-impact decisions. They may include decreases in the time and money it takes to reach a final decision (even if it takes more time and money to reach the organization's policy decision); improved credibility for the organization; and more widely accepted decisions. In opting for a broader analytic-deliberative process, an organization may be required to accept monetary and other tangible costs to gain nonmonetary and intangible benefits, immediate costs to avoid greater future costs, or administrative costs to gain societal benefits.

We are confident that a conscious and careful application of an analytic-deliberative approach will lead to better risk characterizations and better risk decisions.

Six Cases in Risk Analysis and Characterization

This appendix presents brief accounts of six risk decision processes that illustrate some of the points we make about risk characterization. The examples are diverse in terms of the kinds of risk decisions and decision makers they involve and the points they illustrate about analysis and deliberation in informing risk decisions. We have chosen these six chiefly because they are not readily available in published sources. In describing these cases, we do not imply that any one of them successfully characterized the relevant risks. Indeed, some of these efforts were themselves controversial, a fact that underlines the difficulty of designing an effective analytic-deliberative process for informing risk decisions that are likely to become contentious. We believe, nevertheless, that several of the cases illustrate approaches to risk characterization that responsible agencies might find useful to adapt to suit their situations. The cases are presented in the order in which they are mentioned above: ecosystem management in South Florida, incineration siting in Ohio, regulatory negotiation for a disinfectant by-products rule, siting a power plant in Florida, the California Comparative Risk Project, and future land use for a former nuclear waste site.

APPLICATION OF ECOSYSTEM MANAGEMENT PRINCIPLES FOR THE SUSTAINABILITY OF SOUTH FLORIDA

The US Man and the Biosphere Program (US MAB) Human-Dominated Systems Directorate is conducting a 4-year project on ecosystem management for the sustainability of South Florida ecological and associ-ated societal systems (Harwell et al., in press). Although this effort was not envisioned as a project in risk analysis or risk characterization, it addresses important public decisions in which risks are a significant com-ponent. The project is interesting in terms of risk characterization for several reasons. One is its strong emphasis on problem formulation: the project appears to have changed the dialogue on the future of the South Florida environment by redefining the issues into an ecosystem manage-ment framework. Another is its effort to use an analytic-deliberative process to define policy goals that would in turn generate questions for analysis. A third is the project's use of a diverse group of natural and social scientists to represent the concerns of the spectrum of interested and affected parties.

The project focuses on the essential issues related to achieving eco-logical sustainability for the Greater Everglades and the South Florida region and has involved more than 100 scientists representing academic and government sectors in both the natural and social sciences (Harwell and Long, 1995). The project uses the concept of ecosystem management as the framework for harmonizing and integrating the diverse but mutu-ally dependent sustainability needs of society and the environment. This paradigm was not used in South Florida during years of large-scale ma-nipulations of the environment. Quite the contrary: during this century, South Florida has been managed to satisfy human-centered needs with little regard for the sustainability of the ecosystem. The ecosystem man-agement perspective presumes that this approach must evolve to one that explicitly recognizes the mutual interdependence of society and the envi-ronment.

Ecosystem management is a goal-driven framework that integrates scientific understanding of ecological relationships within societal con-texts and emphasizes the need to protect ecosystems and species of con-cern, manage for ecological fluctuations, and employ core reserve/buffer zones to protect the ecosystem over the long term. Because ecosystem management focuses on human and natural systems at regional scales and across intergenerational time periods, it is inherently integrative and adaptive in nature. The US MAB project adopted and applied a set of generic ecosystem management principles (see box, page 171). These principles emphasize the long-term maintenance and sustainability of biological populations, ecosystem structures, functions, and processes, but also explicitly recognize that humans are an integral part of ecosys-tems. This last point cannot be overemphasized.

The US MAB project seeks to replace the divisiveness created by po-larized viewpoints with a spirit of cooperation that fosters the develop-

Ecosystem Management Principles
Applied in the US MAB Project

Use an ecological approach that would recover and maintain the biological diversity, ecological function, and defining characteristics of natural ecosystems. Management must be based upon the science of ecology and shaped by ecological laws. It therefore considers resource decisions at the level of landscapes.

Recognize that humans are part of ecosystems, and they shape and are shaped by the natural system. Sustainable ecological and societal systems are mutually interdependent. Human society and natural systems interact to establish limits and opportunities for action.

Adopt a management approach that recognizes that ecosystems and institutions are characteristically heterogeneous in time and space. Ecosystems and institutions are complex systems operating over different temporal and spatial scales. Ecosystem, jurisdictional, and political boundaries rarely coincide.

Integrate sustained economic and community activity into the management of ecosystems. Strategies should balance human needs with natural systems across all scales. Human communities need to adjust their uses of natural resources to be compatible with intergenerational sustainability.

Develop a shared vision of desired ecosystem conditions. Because decisions regarding ecosystem management have wide-ranging societal implications, the public should have enhanced opportunities for meaningful participation in the decision-making process.

Provide for ecosystem governance at appropriate ecological and institutional scales. Decisions should be made at neighborhood, local, and regional levels with recognition of the interconnectedness of these decisions.

Use adaptive management to achieve both desired outcomes and new understandings of ecosystem conditions. Adaptive management assumes that ecosystems are inherently unpredictable and require flexible policies. Management decisions entail risks, but monitoring and evaluating actions, and subsequent modification of policies, reduce uncertainty.

Integrate the best available science into the decision making process while continuing to improve the basic scientific understanding of ecosystems. Research priorities should be based on a goal of acquiring new information, targeted at management issues, that reduce uncertainties in the policy-making process.

Implement ecosystem management principles through coordinated government and non-government plans and activities.

Source: Harwell et al., in press.

ment of solutions that are beneficial to both society and ecosystems. The project's operating principles raise several management issues:

- the need for a shared vision for ecosystem use and development—although the interests of every group cannot be accommodated, a shared vision can include access to decision-making and a common perception of broad, long-term goals for the region;
- the positive linkage of the environment to sustained economic development—explicit coupling of environmental and economic security rather than the more commonly held view of competition;
- the imperative for adaptive management—recognizing that each decision is simply the best one that can be made under present understanding and that it can be modified and adjusted as new knowledge is gained and uncertainties are reduced; and
- the need for a system of ecosystem governance suitable for implementing ecosystem- and landscape-level sustainability goals—where the hierarchies and complexities of the natural and the human systems are recognized and are directly coupled in governance just as they are in reality.

The organizers of the project presumed that an ecosystem management approach would require integration of theory and knowledge from the natural sciences with analyses of societal and ecological costs and benefits of ecosystem restoration. It would require several kinds of analysis: to identify the defining physical, chemical, and ecological characteristics of the natural, unperturbed Greater Everglades; to use these defining characteristics to develop ecological sustainability goals for the ecosystems of importance in the Greater Everglades; to select methods and ecological characteristics (called ecological endpoints) for assessing and monitoring change; to evaluate the patterns of human uses of environmental resources (such as land and water) and identify other human-caused stresses; to examine the societal and institutional factors influencing ecological sustainability; and to assess the compatibility of ecosystem management with societal policies and institutions and the ability of these policies and institutions to achieve ecological sustainability goals. And it would require deliberation to identify the concerns of the region's interested and affected parties and the social and economic outcomes or endpoints that the project's analyses would have to address.

All potentially interested and affected parties did not participate directly, but efforts were made to have their concerns represented by including in the project a number of experts (primarily social scientists) who were sensitive to those concerns and outcomes because of having worked in and with local groups.

Application to South Florida

The Everglades of South Florida are unique in the world, originally spanning vast open spaces between the coastal ridges of Florida, covering a total area of about 20,000 square kilometers (Bottcher and Izuno, 1994). Wading birds, alligators, sawgrass plains, mangroves, and tropical forests are among the region's most recognizable features, but the essence of the Everglades is the abundance and diversity of species that once lived among the diverse range of habitats (Douglas, 1947; Davis and Ogden, 1994). This "river of grass" flowed from Lake Okeechobee, through sawgrass, hardwood hammock, and pineland communities, to the estuaries of the southern tip of the peninsula (see Figure A-1). The defining features of the natural Everglades consist of the large spatial scale of the system, the highly variable seasonal and interannual patterns of water storage and sheet flow across the landscape, and the very low levels of nutrients in the surface waters. These characteristics led to a unique assemblage of wading birds, large vertebrates, and fish and plant communities patterned in a mosaic of habitats over the landscape and seascape of the region (Davis and Ogden, 1994).

Since the early 1900s the regional environment has undergone extensive habitat degradation associated with hydrological alterations by humans. These were made initially to drain land for agriculture and human settlements, and somewhat later to protect against flooding (Light and Dineen, 1994). The resultant Central and South Florida Project of the U.S. Army Corps of Engineers has created one of the most massive engineered hydrological systems in the world. Additionally, the human population of South Florida is now 4.5 million and growing at a rate of almost 1 million per decade, mostly perched on the narrow coastal ridges. As a consequence of these changes, only half of the original Everglades remains, and only a mere 20 percent of the ecosystem falls within the protective boundaries of the Everglades National Park. The ecosystem continues to degrade, and ecological sustainability cannot be achieved without fundamental changes (Davis and Ogden, 1994; US Man and the Biosphere Program Human-Dominated Systems Directorate, 1994).

The US MAB project included nine steps (Harwell and Long, 1995):

(1) define the geographical boundaries of the regional ecological system;

(2) identify the types of ecosystems that exist within those boundaries and that are of management concern to humans;

(3) identify the natural and anthropogenic stressors on the regional system, including their spatial and temporal components;

(4) identify ecological endpoints for each ecosystem type, where eco-

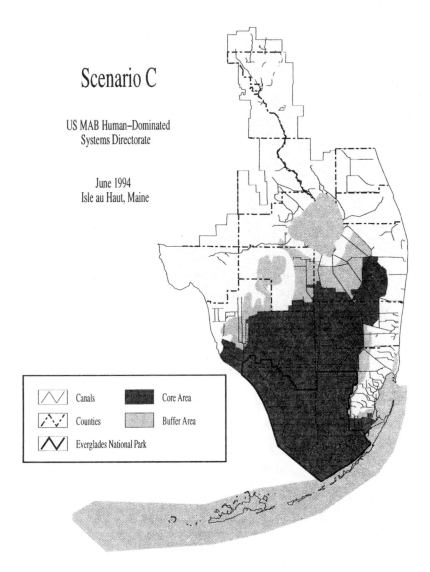

FIGURE A-1   A scenario for management of the greater Everglades ecosystem.
SOURCE: Harwell, Long, Bartuska et al., in press.

logical endpoints are defined as the ecological characteristics across a
range of hierarchy (population, community, ecosystem, and landscape
levels) that can be used to evaluate the health or change-of-health of the
ecosystem;

(5) specify the ecological and hydrological characteristics of a sustainable ecosystem, defined in terms of the selected ecological endpoints;

(6) characterize the human factors affecting the ecosystem, including stressors, feedbacks to society, and societal values of the ecosystems;

(7) define ecological sustainability goals for each component of the landscape, with focus on core areas of maximal ecological goals and buffer areas to support the attainment of those goals;

(8) establish plausible scenarios of management of the regional system; and

(9) examine those scenarios for their implications for the desired goals for sustainability of the regional ecological and societal systems.

The study presumed that sustainability for the South Florida regional ecosystem would require the reestablishment of enough of the natural hydrological system to provide water quantity, timing, and distribution over a sufficiently large area to support the ecological components, such as wading birds and the mosaic of habitats, that constitute the essence and uniqueness of the Everglades (Harwell et al., in press). The study concluded that the environment of South Florida has more than sufficient water except in severe drought years to support all anticipated urban, agricultural, and ecological needs, but that the major portion of that freshwater is lost directly to the sea through the engineered system of drainage canals. The critical issue then is not competition for resources, but the storage and wise management of this renewable resource.

The US MAB project used the scenario-consequence analytical approach to examine environmental effects of human actions. This approach involves developing a hypothetical set of conditions (scenarios) that are internally consistent and scientifically defensible and that specify all important factors needed to evaluate effects (Harwell et al., 1989). A scenario is meant to be neither a prediction of the future nor a proposed plan of action; rather, it is meant to cover the range of situations that are sufficiently plausible to warrant further evaluation. The relative risks, costs, and benefits of the plausible management strategies can then be evaluated comparatively, providing a much stronger basis for selecting among options.

The US MAB examined three scenarios in terms of how associated changes in land use and hydrology would affect the defining characteristics of the natural Everglades: spatial scale, dynamic storage and sheet flow, and habitat heterogeneity. Scenarios were aggregated according to land use designations:

Scenario A: management involving only existing publicly held wetlands in the Everglades basin;

Scenario B: the addition of contiguous, privately owned areas within the original Everglades that remain as functional wetlands;

Scenario C: the further addition of contiguous, privately owned areas that no longer are functional wetlands, but that could provide water storage and management functions or could be recovered as functional wetlands.

Land use was classified into core and buffer areas: core areas are the wetlands that would be managed to recover predrainage hydrological patterns in terms of water quantity, distribution, and timing (where maximum recovery of the defining characteristics of the natural ecosystem is assumed to occur); buffer areas are parts of the system to be used for water supply for both ecological and human needs, flood protection for urban and agricultural areas, enhancement of water quality, and as hydrological transition zones between natural conditions and managed areas. Variations in scenarios were developed that differed in the location and extent of core and buffer areas. All scenarios presumed flood protection and other support required for the human-occupied zone.

It was concluded that Scenarios A and B contain insufficient spatial extent of the core area to provide for the defining characteristics of the Everglades at population to landscape levels or contain insufficient buffer area in order to provide for hydrological storage and release similar to natural hydrological cycles. That is, neither scenario was considered ecologically sustainable.

Scenario C (see Figure A-1) involved using portions of the Everglades Agricultural Area (EAA) for dynamic water storage while it remains entirely or in part under private ownership. The EAA presently consists of 280,000 hectares, primarily under sugar production, with total annual economic activity of approximately $1.2 billion (Bottcher and Izuno, 1994). Scenario C was considered sufficient to achieve the ecological goals for the core area, but variations in the scenario relating to the amount of EAA lands that would be acquired publicly and therefore taken out of agricultural production had major societal implications.

Complete acquisition of the EAA was concluded to have too high an economic and social cost for the communities of this historical agricultural area (Bottcher and Izuno, 1994). Yet the sustainability of the sugar industry in the EAA itself is at risk for several reasons: extensive soil degradation, which has been caused by the lowering of the water table and extensive microbial oxidation and loss of the peat soils; potential changes in sugar price supports because of liberalization of international trade; likely political changes relating to Cuba that would affect current import bans on sugar; political forces aligned against sugar production in the EAA, including efforts to tax the sugar industry exclusively for funds to restore the Everglades; and economic pressures to acquire EAA lands for residential development.

Consequently, bringing at least part of the EAA into a buffer function (water storage and management) in support of the ecological systems of the region might counteract the risks to the sustainability of the agricultural system. The US MAB scenario suggested possible uses for the EAA that would allow for sugar production to continue and for the water management needs to be met, thereby linking the sustainability of the ecological system with the societal sustainability of the local community. One possibility is the development of sugar cultivars that would be highly productive under flooded conditions; another is the creation of economic incentives for water storage by sugar farmers, such as subsidizing flooded-system sugar prices or paying farmers to store water rather than grow crops. An interesting result of the analyses is that if any agriculture is to remain in the EAA, sugar is probably the most desirable ecologically, as its nutrient demands and nutrient exports to the Everglades are perhaps an order of magnitude lower than those of vegetable crops. Furthermore, sugar agriculture is much preferable to the alternative of housing developments in terms of the impact on the Everglades system downstream.

Further detailed analyses remain to be done, but the risk characterization using an ecosystem management framework in the US MAB project has now suggested that a solution may be feasible that achieves ecological sustainability of the regional ecosystem and is consistent with the economic, cultural, and other societal sustainability goals for the agricultural community of the region.

Implications of the Case Study

The ecosystem management case study in South Florida illustrates some points that have more general importance for environmental risk characterization. First, it illustrates a way to address seemingly more complex issues than traditionally addressed by risk assessments (i.e., single chemical, single health effect) by examining the many outcomes of a few plausible scenarios as a way of understanding the risk situation.

Second, it illustrates the critical role of problem formulation in the ecological risk assessment paradigm (U.S. Environmental Protection Agency, 1992a) and its constituent elements. These elements include identifying the at-risk systems or populations; selecting the full range of outcomes to those at-risk systems or populations that must be characterized; identifying the types of information, analyses, analytical methodologies, and other tools needed to characterize the risk situation; making explicit the contextual issues and their implications; and mapping the problem onto a risk characterization landscape to provide guidance to the decision makers.

Third, it illustrates the use of an adaptive management framework for risk decision making and the place of risk characterization within that approach. Adaptive management is fundamentally a problem-driven approach. It begins with explicit objectives (in this case, ecological sustainability that is consistent with societal sustainability); takes a long-term perspective, recognizes that long-term achievement of environmental goals affects and is affected by the societal context; and adopts a policy strategy of making interim decisions, monitoring consequences, and altering the decisions as conditions warrant. It relies on analyses that are interdisciplinary (especially across boundaries between natural and social sciences) and that focus on reducing uncertainties. It also relies on deliberation, beginning with efforts to develop a shared vision or problem formulation that might be accepted by many affected parties. In the South Florida case, that effort began only after the initial analytical stages of developing options and scenarios, and it involved participation by analytical experts with varying disciplinary perspectives and familiarity with the parties' perspectives, although not by the parties themselves. It is too early to tell whether a shared vision will be widely accepted by the people of South Florida. The key result of the analysis and deliberation was a set of scenarios and consequence analyses that essentially characterized the risks and other outcomes of various plausible management strategies and served as input from the program to the decision participants.

The adaptive management framework, as applied in South Florida, faces limitations involving such issues as data needs, the potential for adversarial processes to interfere or derail the process, the perception that it is controlled by elites, and the time required for adequate development of the interdisciplinary research team. The approach is geared for addressing large and complex problems, those that are multidimensional and cover a long time frame or large spatial scale.

APPROVAL OF THE WASTE TECHNOLOGIES, INC. INCINERATOR AT EAST LIVERPOOL, OHIO

Controversy surrounding the Waste Technologies Industries (WTI) hazardous waste incinerator, located in East Liverpool, Ohio, reflects a number of issues common to hazardous waste facility siting, and particularly incinerators. Planning for the incinerator on a 20-acre plot of land next to the Ohio River, previously owned by the county port authority and zoned for heavy industrial development, began in 1980. A long history of permit applications, challenges, appeals, disputed approvals, and legal maneuvering preceded construction work from 1990 to 1992, and assessment and dispute has continued since then. A series of air pollution exposure and risk studies were conducted during the early planning period, with a new formal effort for risk assessment initiated by the U.S. Environmental Protection Agency (EPA) in 1992 and continuing with review and updates to the present. These studies have been supplemented by a series of controversial test burns that began at the end of 1992. The site has been the subject of much scientific, technical, and political debate on the safety and appropriateness of incineration, capturing the attention of the national media and producing impacts on national EPA policy and presidential politics.

The EPA efforts at risk assessment of the WTI facility initially focused on the cancer risks to the population and to the hypothetical maximally exposed individual that would be associated with permitted, routine air releases from the incinerator stack and subsequent exposure through inhalation and ingestion of water and locally grown food, including meat and dairy products. Contaminants of concern included dioxin and other stack emissions. More recently, following review, the risk assessment was broadened to include health, safety, and ecological concerns, including releases during startup, shutdown and malfunction/upset conditions, and on-site and off-site transportation accident risks (U.S. Environmental Protection Agency, 1993a; A.T. Kearney, Inc., 1993; Johnson, 1996). Draft results of the risk assessment, still undergoing peer review, suggest that cancer risks are less than 1 in 10^6 for a nearby resident, and that noncancer health risks and risks from accidents and ecological impact are low and acceptable (Johnson, 1996). These results, and the belief by many that the WTI facility is "world class," with the best state-of-the-art pollution control equipment available, form the basis for arguments by proponents that continued operation will be both safe and profitable. Yet significant controversy remains.

For many, the debate over the specifics of the WTI risk assessment has served only as a surrogate for continuing societal concerns over broader issues, including the role and suitability of incineration as a hazardous waste management option. The uncertainy and controversy surrounding dioxin has also affected the WTI risk debate, as it has at other incinerator sites. Beyond this, some people believe that approving the operation of any hazardous waste management technology provides an incentive for continued production and distribution of chlorinated or other potentially hazardous compounds. These systematic concerns ebb and flow into any local debate over incineration, but they are often pushed off the screen in formal risk assessment and risk characterizations for a proposed project.

The current risk assessment efforts are for a facility already built and undergoing test burns. Indeed, proponents of incineration maintain that this is the only way to conduct an effective risk assessment: site-specific burn efficiencies and operating parameters are needed to effectively evaluate emission rates. However, opponents argue that such a procedure, allowing extensive costs to be sunk into the construction and testing of a facility prior to the risk assessment, virtually ensures that approval will be granted. Effective and cooperative stakeholder participation in the risk analysis process is indeed difficult when some participants feel that the decision has already been made—that it is "a done deal"—and that the results of the risk analysis must perforce conform to this decision. This timing virtually assures that many will view the risk assessment as inherently biased or irrelevant.

A number of participants in the WTI debate have noted the need for inclusion of a broader range of considerations in the risk evaluations. In particular, they have argued that the assessment of risks to a population due to a single source cannot be conducted absent the context of other risks imposed on that population and that particular subgroups within the population may be subjected to an especially high combination of exposures with implications for both health risk calculations and environmental equity concerns. These issues are often raised as part of the broader agenda of environmental groups in debates over hazardous waste facilities. In public comments during the December 1993 EPA review meeting of the planned risk assessment for WTI, Rick Hind, representing Greenpeace, posed the following questions (U.S. Environmental Protection Agency, 1993a:Appendix H, Observer Technical Questions):

> When assessing risks to sensitive subpopulations near WTI, how will you consider the effects and sensitivity (body burden) on the African American community closest to WTI? How will you address WTI risk posed to these and other children living and attending school (nearby to the site)? Will you look at body burden already found in these subpopulations from other sources including possible immune suppression from toxic chemical pollution and radiation?

East Liverpool resident Marilyn Parkes further noted the need to consider the WTI emissions in the context of other industrial exposure sources, including those from an area power company that had purchased excess emission permits from another region. These requests for more detailed analysis raise legitimate issues. For some of those who raise them, however, doing so may serve to advance in-principle objections to incineration that are allowed no legitimate place in the discussion when the problem is formulated in terms of risk assessment for a particular incinerator (see Jasanoff, 1986).

Another strong current in the WTI debate is the need to address local public health concerns by monitoring measurable environmental outcomes in the community, rather than using only theoretical, model-based risk predictions. In a letter dated December 6, 1993, to the EPA Region 5 Administrator, the City of East Liverpool Board of Health states (U.S. Environmental Protection Agency, 1993a:Appendix J):

> The only real data which EPA proposes to use in the Phase II Risk Assessment involve trial burn emission data. This is not sufficient to gain public confidence in East Liverpool. . . . Community monitoring programs need to be expanded in scope and frequency. This must include monitoring of air, soil and crops (i.e. foodchain) at numerous sites within the community. . . .
>
> Any assessment that involves only theoretical assumptions on community exposures stands little chance of gaining public support. The East Liverpool Board of Health advocates the development of an expanded and accelerated emission and environmental monitoring program rather than a quantitative risk assessment based on little real data. The East Liverpool Board of Health encourages cooperation with Federal EPA in further development of specific data collection programs which can provide a realistic basis for protecting public health.

These comments illustrate the difference between a public health paradigm (Burke, 1995) and that implicit in the current prevailing approach to scientific risk assessment. Such concerns have remained despite the fact that the scientific studies in support of the risk analysis were detailed and extensive. Indeed, the extent of mistrust in risk assessment and those performing it has been indicated by local residents' suggesting "the possibility of scientists 'fudging' risk assessment data" (Johnson, 1996). The long-term impacts of continued operation of the WTI incinerator may not be known for many years. Assuming the results of the current risk assessments are representative and correct, the facility could provide a safe, effective, and economically beneficial mechanism for hazardous waste management. It is clear, however, that the concerns of all participants have not as yet been addressed by these assessments, and for many, the full dimensions of risk from the WTI facility have yet to be characterized.

REGULATORY NEGOTIATION FOR A
DISINFECTANT BY-PRODUCTS RULE

Regulatory negotiation is a process in which representatives of formal stakeholder groups work consensually with government regulatory bodies to draft a proposed rule (see Appendix B for more detail). Regulating negotiation can provide a way in which analysis and deliberation can be coordinated to inform risk decisions. Without endorsing the specific process used in this case study, we present it in detail as an illustration of how respect for a few very simple principles can help characterize risks in a way that meets the needs of public officials and the interested and affected parties.

Chlorination of drinking water is the most common technique used by water suppliers to reduce the risk of microbial infectious pathogens. However, although chlorine and other disinfectants do reduce microbial risks, they also react with organic compounds already present in the water—disinfectant by-products (DBPs)—some of which are carcinogens. It is these by-products that EPA wanted to regulate. Under the Safe Drinking Water Act, EPA was obliged to set standards for drinking water contaminants. The proposed DBP rule represents a portion of this effort.

EPA turned to regulatory negotiation because any DBP regulation would have been contentious using the customary rule-making process, for two main reasons. First, EPA had previously concluded that the available data were inadequate to address concerns about DBP and microbial risks associated with disinfection of drinking water. Any proposed rule would have to invoke judgments that EPA did not believe it could adequately substantiate with evidence. Second, the Safe Drinking Water Act was soon to come before Congress for reauthorization. The water supply industry was already on a collision course with environmentalists and consumer advocates. The water industry hoped to gain relief from the additional—and, in its view, unreasonable—regulatory burdens demanded in the act and environmentalists wanted to preserve the act's central mandate. Meanwhile, EPA had already entered into a consent order with a litigant, obliging itself to propose a DBP rule by June 1994. Regulatory negotiation offered a way to bring would-be adversaries together, to give them a chance to speak with each other, to work with EPA to devise a proposed rule, and to give all parties involved a chance to recognize that the conflict over drinking water regulation was not likely to be won by any single party, but, instead, might be resolved by cooperation and compromise.

In September 1992 EPA announced its intention to organize a regulatory negotiation rule-making process to develop a DBP rule. Between November 1992 and June 1993 a negotiating committee met eight times, reviewed evidence, ordered analyses, negotiated, and developed proposed rules that all participants might accept. It consolidated 16 policy options to 3 and then asked a technical assistance group to perform cost-benefit analyses and compliance assessments for the 3 options. The negotiating committee resolved disagreements and issues of uncertainty in creative and unexpected ways. For instance, instead of producing a single DBP proposed rule, the committee proposed three rules. Only one addressed DBP regulation directly. Furthermore, the proposed rules made allowances for not burdening small water supply systems with complex and costly measurements and treatment technologies by arranging for regulations to be phased in over time, and dealt with the issue of information gaps by proposing that DBP rules be refined as more data becomes available.

Forming the Negotiating Committee and Technical Support

Under the Negotiated Rulemaking Act of 1990, certain criteria are used to establish if regulatory negotiation would be productive and worthwhile. A main objective is to determine how feasible it is to compose a negotiating committee that adequately represents the full spectrum of interests on the issue. In this case, EPA hired an outside firm, which conducted more than 40 interviews with agencies, water suppliers, environmentalists, equipment manufacturers, and consumer groups. From these interviews it was established that the number of affected interest positions was relatively small, the factual base for holding deliberations was well developed, there was a strong degree of "good faith interest" in resolving the issue through negotiation, and the agency was willing to commit the necessary resources. These results satisfied the agency's criteria for initiating regulatory negotiation.

On the basis of the preliminary interviews, EPA proposed a negotiating committee of 17 individuals. After public review, one more member was added. A dispute arose about whether or not chlorine chemical and treatment equipment suppliers deserved a seat at the table. Clearly these stakeholders had valuable knowledge and advice to offer, but they were also strongly committed to the use of chlorine as the primary drinking water disinfectant. A creative compromise was struck. The chlorine industry would not be granted a seat on the negotiating committee, but would instead participate in an advisory role by serving on the Technologies Working Group (TWG). A representative was invited to state the group's position at the very first negotiating session.

The TWG provided a formal opportunity for chemical and equipment suppliers who had not been named to the negotiating committee to contribute input to the negotiations, while also supplying much-needed information about the cost and performance of drinking water treatment technologies. In addition, EPA arranged for three experts to provide ongoing scientific advice and technical support for those members of the negotiating committee whose organizations were not able to provide this kind of expertise.

Defining the Problem and Informational Needs

In all, six negotiating meetings were planned. The general strategy was to follow the following sequential steps: gather information, develop evaluative criteria, generate options, evaluate options, draft agreement, and obtain closure. In practice, the process was much more iterative and integrated. For example, the committee defined criteria at the first and second meetings, gathered information throughout the entire process, and reached closure on some items earlier than others. It took eight meetings to reach partial closure. Over the next 8 months, participants did not meet, but drafted rules, reviewed them, and finally gave approval.

EPA initially framed the problem in a 1991 EPA Status Report (59FR 38675) that mentioned the need to regulate DBPs by assigning maximum contaminant levels for individual compounds or for groups of compounds. It also framed the inherent problem as a risk-risk tradeoff. Major difficulties were expected to be the reduction of DBP risk without a simultaneous increase in microbial risk and the introduction of new or unknown risks from changes in treatment technology. In the status report, EPA clearly stated a position against adoption of a rule that would promote a shift to non-chlorine disinfectant technologies. The report suggested recommending continued use of chlorine disinfectant by specifying that as the preferred choice of technology—best available technology or BAT.

When the negotiations began, not all members of the negotiating committee shared the opinions expressed in the 1991 status report. There were substantially different perceptions of the problem among members of the committee, including differing perceptions of the risks associated with DBPs. Many members expressed the opinion that available scientific and technical information was inadequate to justify proposing specific rules. Some members believed that only slight improvements in disinfectant technologies were justifiable until more became known about the relative risks. Others believed that it would be best to specify treatment techniques and not specific contaminant levels. Still others advocated a radically new approach of regulating DBP precursors, namely the elimination of organic material before treatment through a combination of filtering and enhanced watershed protection.

The negotiating committee was not free to define the problem as it wished because it was constrained by law. The Safe Drinking Water Act requires that contaminants be regulated with maximum concentration levels. Treatment techniques can be the basis of a regulation, but only when it is not economically or technologically feasible to measure the contaminant. Watershed protection was an attractive target for the committee, but it is covered under a different rule (the enhanced surface water treatment rule). The committee frequently found these constraints to be barriers to a creative agreement and did not always accept them unquestioningly. In fact, the committee pursued several alternatives that seemed to run contrary to the letter of the law. For example, the committee asserted that DBP rulemaking should be consistent with other rules, and it proposed a new enhanced surface water treatment rule as well as the DBP rule.

The negotiating committee did not begin by attempting to define the problem. Instead, it sought to define the characteristics of a good solution. One of the first tasks was to define the value objectives of a "good" rule. In facilitated open discussions, the committee produced an unranked listing of value objectives to be considered in the decision-making process. The list included whether a potential rule was protective of human health, flexible to source water quality and existing treatment facilities, protective of environmental equity, sensitive to needs of susceptible populations, consistent with EPA rules, explainable to the public, and affordable. These criteria would be used in later sessions to evaluate competing proposals.

At the first formal negotiating session, most participants agreed that some type of DBP rule was needed. There was less acceptance of EPA's definition of the problem as a need for maximum contaminant levels. Some believed that the committee should consider a pollution prevention approach that aimed at eliminating the DBP precursors. (Precursors are naturally occuring humic and fulvic acids that react with disinfectants to produce dangerous DBPs.) Theoretically, most of the DBP problem would disappear if precursors could be removed before treatment. Precursors can be reduced though watershed protection measures (which bring other additional benefits), granular activated carbon, membrane filtration, or enhanced coagulation. The participants who defined the problem in this manner advocated the development of a rule to enhance the existing surface water treatment rule (which is basically a watershed protection rule).

The committee early recognized the need for more data and analysis to inform its discussion of the competing problem formulations. To address these needs, the committee consulted with the TWG in order to specify its informational needs. A technical workshop was organized for the purpose of informing negotiating committee members on the range of scientific opinions about health risks, treatment technologies, costs, and modeling efforts. Twenty-three nationally recognized experts on drinking water treatment gave presentations and participated in panel discussions for the benefit of the negotiating committee. As questions arose during the following months, additional presentations or testimony were given to the committee.

Integrating Deliberation and Analysis

Throughout the negotiations, the negotiating committee requested studies, evaluations, data, and advice from the TWG. At the very first negotiating session, for example, the TWG was asked to organize the available information about treatment technologies. One participant suggested the results be presented as a matrix comparing treatment alternatives along several performance criteria. These criteria were defined by the negotiating committee only in the vaguest sense. It remained for the TWG to specify the criteria, to develop indicators, and to make assessments on the measures. Three weeks later, at the second negotiating session, the TWG presented a schema of 31 treatment scenarios, a standardized coding form, and results for 10 completed evaluations.

Negotiating sessions usually began with members of the TWG responding to requests from the previous session. Often, these exchanges raised more questions than they answered and resulted in more tasks being assigned to the TWG. For example, after receiving the analysis of cost to each household, the negotiating committee then asked the TWG to estimate the national costs of each treatment technology. The TWG responded that that question could not be answered without first gathering more baseline information about the quality of pretreated water across the nation. This need for more baseline information would arise repeatedly throughout the negotiations.

In some instances, the TWG was able to provide clarification that helped resolve misunderstandings or disagreements among the negotiating committee. One definitional issue that arose concerned a technology known as granular activated carbon. Different committee members had different understandings of what the technology was and what it was capable of accomplishing. A TWG member cleared up the misunderstanding by explaining that there were basically two major types of granular activated carbon systems. Later, the negotiating committee would ask the TWG to write an official definition of the two types of systems to resolve the confusion. The TWG also performed risk analyses and cost-benefit and other economic analyses at the request of the negotiating committee. At times, members of the committee were assigned to work with the TWG in these analyses.

At the third negotiating session, the committee asked the TWG to develop information on the distribution of source water characteristics across the nation. This information was needed to satisfy the objective of environmental equity, giving each community the same level of risk protection. Rules that require 99.99 percent reduction of viruses do not provide equivalent risk protection when source contamination varies over several magnitudes. The negotiating committee wanted to know how much variation in source water contamination existed and how that was distributed among different communities.

Most of the TWGs effort went into computing cost and compliance figures for the different approaches. Sensitivity analyses were also requested. These data and analyses were needed to meet the stated objective of an affordable rule. Cost analyses that were originally ordered in January were presented at the next meeting in February and revised, and more sophisticated results were presented at later meetings. This cycle of request/report/revise/report was characteristic of this regulatory negotiation process.

Information and results of analyses from the TWG were also used by the negotiating committee to help resolve fundamental differences in regulatory approaches. For example, by the third session, the committee was considering three basic approaches. A pollution prevention approach involved reducing the entry of DBP precursors into the water treatment system or removing them before the disinfectant was added. In order to investigate the hypothesis that precursor treatment could provide adequate risk protection, the committee asked the TWG to analyze the effect of precursor content on DBP levels in treated water. The TWG had to address the question of which variables to measure in the untreated water.

A Creative Solution to Handling Uncertainty

During the deliberations, the committee determined that there was still insufficient information available to make many appropriate decisions. Specifically, it needed more field information about the existing nature of risks from microbial pathogens and the capabilities of different types of treatment technologies to control those pathogens. The negotiating committee responded to this need by preparing, with the assistance of the TWG, an Information Collection Rule (59 FR 6332 [February 10, 1994]).

The proposed rule requires large public water supply systems to monitor source water for microorganisms and viruses, test for the presence of DBP precursors before and after treatment, test for DBPs, and provide engineering information about treatment techniques. In some instances, systems reporting high levels of precursors would be required to conduct special pilot tests to reduce them. This amounts to a pretest of the field performance of treatment technologies being considered for the DBP rule. The purpose of the pilot testing is to obtain more information about cost effectiveness of precursor control technologies. Language in the proposed rule specifies minute technical details referring to the description of acceptable pilot technologies, their implementation, and testing and reporting protocols.   •

The Information Collection Rule represents the breakthrough compromise that made the proposed DBP rule politically feasible. Although there was wide agreement at the very start of the process that a DBP rule was needed, informational gaps stood in the way of resolution and final agreement. One thing the committee did agree on was the need to reduce uncertainties by gathering more data. Data collected under the rule are to be used to inform the final adjustments of the proposed rules and the need for and content of long-term rules.

Rather than put off the entire rule-making process until after all relevant information had been collected, the negotiating committee decided to agree on provisional rules with the understanding that a second stage of negotiations would be necessary some few years in the future. Stage 2 rules would be revised on the basis of the data gathered in the meantime. Committee members advocating stricter rules demanded that Stage 2 "straw man" provisions be set, and set intentionally low in order to ensure that all committee members would return for those negotiations. In addition the committee announced it would meet again should additional information become available on acute health effects that would necessitate more expeditious changes in rules.

Conclusions

The DBP regulatory negotiation represents a creative approach for dealing with conflict and uncertainty by reducing uncertainty over time. It also illustrates how regulatory decisions can benefit from processes that promote meaningful deliberation among experts, public officials, and interested and affected parties. A variety of deliberation techniques were used to help build mutual understanding and consensus among the negotiating committee members. Deliberation was integrated with analysis through requests to and reports from a technical advisory committee.

The DBP regulatory negotiation was a success in the sense that it resulted in the publication of three proposed rules, including the DBP rule, that are in the process of final review. Before the process began, there was considerable antagonism among the stakeholders in the process. EPA was clearly obligated to propose a rule, but it expected that any such rule would be challenged in the courts. Meanwhile, the Safe Drinking Water Act was scheduled to be reauthorized by Congress, and it appeared possible that Congress would entertain proposals to alter the act. Several parties were hopeful that Congress would amend it so as to not require such demanding regulation. For all these reasons, the general expectation was that a DBP rule, if passed, would probably not be very comprehensive.

The negotiation also led to a good-faith effort to cooperate in funding research into disinfectant by-products and their risks. The American Water Works Association Research Foundation committed several million dollars, which was matched by EPA and further supplemented by Congress. The consensus is that this money and cooperation will bring about a lot of good, relevant research that otherwise would never have been done.

One member of the negotiating committee, the representative of small, mostly rural water suppliers, did drop out of the process when he decided that he would be able to protect his group's interests better by acting outside the negotiation. He concluded that the group's concerns for high costs of new regulation could best be served by appealing directly to Congress. This group never acknowledged there was a need for DBP regulation. Once its proposals were rejected by the other committee members, its representative withdrew. The groups efforts to influence Congress apparently bore fruit in 1994 in special amendments in the Senate to the Safe Drinking Water Reauthorization that eased the regulatory burden on small water suppliers. (The bill did not clear the conference process because of unrelated issues.) As of early 1996, the small water suppliers had not commented on the proposed rules or challenged them in court.

The disinfectant by-products regulatory negotiation provides an example of how analysis and deliberation can be coordinated in making risk decisions. Throughout the negotiations, committee members repeatedly drew upon technical advisors to interpret and analyze data, to model risk scenarios, and to estimate costs and feasibility of regulatory options. Three positive aspects of this case are worth reiterating: the use of a technical advisory committee and expert assistance; the manner in which the negotiating committee dealt with the issue of data inadaquacies; and the iterative nature of the negotiations and rule-making process. In addition, the process proved widely acceptable to almost all of the interested and affected parties (U.S. Environmental Protection Agency, 1992c, 1992d, 1992e, 1993b, 1993c, 1993d, 1993e, 1993f, 1993g).[1] (For a more detailed account of the role of analysis in the regulatory negotiation, see Roberson et al., 1995.) It is also worth emphasizing one negative aspect: the possibility that a single party may be able to trump the entire process of analysis and deliberation with an intervention in the political arena. This pos-

---

[1]See also Federal Register 59(28):6332-6444 (Feb. 10, 1994); 59(145):38668-38829 (July 29, 1994); 59(145):38832-38858 (July 29, 1994).

sibility casts a shadow over the entire process of risk characterization and decision making in government agencies. It is especially significant for approaches that make special efforts to broaden participation, because the success of an interest group in going outside a deliberative process can increase alienation and mistrust on the part of participants who were coaxed, perhaps reluctantly, to participate in a process they did not initially fully trust.

SITING A POWER PLANT WITH THE AID OF DECISION ANALYSIS TOOLS

In the late 1980s, the Florida Power Corporation (FPC) began a search for one preferred and two alternate sites for a 3,000 megawatt coal-fired power plant. Knowing that public scrutiny would be keen, FPC decided to avoid the traditional failings of a decide-announce-defend approach to siting by drawing multiple perspectives into the site selection process early on. Two groups—a committee made up of FPC managers and staff and an Environmental Advisory Group (EAG) composed of eight high-profile figures in Florida—engaged in a decision-making process that employed multiattribute utility analysis to help make the issues transparent and logical and to focus the debate in a constructive manner. By all apparent measures, the process was successful. The EAG had clear and specific influence in the site selection process, and licensing is under way without opposition at the selected site.

Multiattribute Utility Analysis

Multiattribute utility analysis is a technique for formally drawing multiple perspectives and evaluations into a decision making process. It begins with the working assumption that decisions have multiple goals, not all of which are equally important. Differences are attributed to the particular interests of the interested and affected parties and the decision makers. The technique makes these differences (and similarities) lucid by eliciting from participants their subjective judgments about the importance of outcomes (subjective utilities) and using these as a basis for comparison. The goal is not to reach a forced "consensus" through averaging, but to clarify positions and to test the feasibility of various policy objectives.

A typical approach in multiattribute utility analysis begins by asking each participant or group of participants to list and set priorities among the criteria that they would use to evaluate the decision options. For example, in a siting case, one person may value visibility of the facility from the town center as paramount, while another may emphasize noise or emission levels. With the help of a decision analyst, these criteria are organized into a hierarchy called a "value tree." Value trees can be composed for individuals and for groups. The next step is to assign relative weights to each criterion. As the criteria are arranged hierarchically, each "branch" is given a relative weighting. For example, all criteria having to do with aesthetics might be grouped as one branch of the tree, while another branch may include all criteria concerning emissions. There is no single right way to group or weight the criteria; rather, the goal is to make certain that the result accurately reflects the concerns and judgments of the participants. "Twigs" on each branch are indicators or measures of how well a decision option performs on that criterion. Overall scores for decision options from the perspective of a participant can be computed by first applying the indicators and then summing each weighted branch using simple algebra.

Multiattribute utility analysis requires coordination between deliberation and analysis at two levels. First, the computation of scores, an analytic procedure, follows deliberation in which values are elicited, discussed, peer-tested, and revised. Deliberation plays a major role when this analysis deals with groups of people rather than individuals. For example, in the Florida Power Corporation case, two different groups did separate analyses, and in each group, the members deliberated to reach agreement on the evaluative criteria and their relative weightings. The different results allowed for further interaction and learning.

Analysis and deliberation are also intertwined in quantification. Depending on the nature of the indicators and the available data, measures may be made by technical procedures (such as measuring the permeability of subsurface soils), or quantification that may be highly subjective (such as nuisance from additional traffic). Analysis provides the former measures; deliberation arrives at ways to quantify factors that are not readily measured. Deliberation is also important in reading and interpreting numbers that combine such disparate kinds of indicators.

The Decision Process

Florida Power Corporation was assisted by consultants specializing in decision analysis techniques. Their task was to elicit evaluative criteria from the FPC group and the EAG, operating separately, and then to help the groups come to consensus on how to weight the criteria relative to each other. This was accomplished in a step-wise manner, beginning with a large search area and a few general criteria and working toward more specific criteria and a narrower search area. In between EAG meetings, the consultants would apply the criteria to the potential sites. The results were presented at the following meeting. Exclusionary and screening criteria were used to eliminate unworthy sites from future consideration. Potentially suitable sites were ranked using each group's weighted criteria. Results from the FPC group were shared with the EAG and vice versa. As a result, the FPC group reconsidered certain weightings, and the two groups' judgments came closer together as the project progressed. The final site chosen was highly recommended by both groups.

The siting study was conducted in five phases and took more than 18 months to complete. Each phase was based on successively more refined data and criteria. Input from the EAG was used only in Phases II, IV, and V, due to limitations on the amount of work the EAG was able to undertake. It is in these phases where the interplay of analysis and deliberation can be seen most clearly.

The EAG, as noted above, was made of eight high-profile individuals, including local and state government officials, academics, an environmental attorney, and representatives of environmental and business groups. Its purpose was to bring multiple values and perspectives into the siting process early on. This approach demanded significant contributions from the sponsoring company and the individuals who agreed to serve on the EAG. Over a year's time, the EAG convened regularly for all-day meetings.

Ideal MUA procedure would be to elicit the evaluative criteria from the EAG and the FPC group from scratch. Because of time constraints the consultants, working with staff from FPC, developed a preliminary set of working criteria and measurement indicators. These were used as a starting point for the EAG. Using a structured elicitation protocol, the consultants' first step in Phase II was to have the EAG review and modify the preliminary list to better suit their needs.

Phase II proceeded in six steps:

1. The consultants elicited a set of criteria and weights from the FPC panel.

2. Using the FPC criteria (though not its weights) as a starting point, a set of criteria and weights was elicited from the EAG; see Table A-1. The EAG supplemented its list with justifications and explanations.

3. The FPC group reviewed the results from the EAG and modified its criteria set to match the EAG set. As Table A-1 shows, it also modified many weights in the direction of the EAG weights. Wetlands disruption was revised from a weight of 54 to 14, very close to the EAG's value of 15. Based on argument given by the EAG, the FPC group decided to prefer sites that were already near a disturbed area, and they added this new criterion to their list. The FPC group then increased weighting on several other criteria (effects on county tax base, disturbed areas, sensitive land uses, surface water, and terrestrial communities), bringing them closer to

TABLE A-1   Criteria and Weighting for Power Plant Site Selection

Criteria	FPC Initial Weight	EAG Weight	FPC Final Weight
Impact on Surface Water Quality	11	15	19
Wetlands Disruption	54	15	14
Proximity to Disturbed Areas	not used	19	11
Proximity to Sensitive Terrestrial Communities	6	15	11
Proximity to Sensitive Land Uses	6	15	11
Proximity to Class I Air Quallty Areas	10	12	11
Impact on County Tax Base	3	10	11
Proximity to Urbanized Areas	6	0	11
Existing Air Sources	6	0	0
Total	102	101	99

the EAG weights. On only one criterion, proximity to urbanized area, did the FPC group's weighting grow further apart from the EAG's.

4. The consultants used the EAG criteria and weights to rank the 172 areas and select 59 sites.

5. The consultants used the FPC criteria and weights to rank the 172 areas and select 57 sites.

6. The consultants combined the two sets of selected sites and set a cutoff threshold in collaboration with FPC staff and after considering graphical plots of the site evaluations. Because of the FPC group's shift in weights, there was agreement on 55 areas sites: 4 were selected only by the EAG and 2 were selected only by the FPC group. All 61 sites were sent to Phase III for further consideration.

The single criterion added by the EAG was quite significant. FPC's set included criteria that selected for sites away from people and areas of development. Fearing that a remote site would promote sprawl, the EAG advocated sites on already "disturbed" land near, but not directly on, existing settlements, which could house and support the plant's employees. After much deliberation, the EAG discovered it could capture these concerns with a criterion called "Proximity to Disturbed Areas." That criterion was the highest weighted by the EAG and the third highest weighted (ultimately) by the FPC. The effect of combining that criterion with the others favoring sites far from people was to select for remote, but not environmentally attractive areas. In fact, the site ultimately selected was just such a site.

In sum, the EAG affected the Phase II ranking in three ways. First, it added an important criterion that FPC had not originally included. Second, it assigned different weights among the criteria than did the FPC group, and the resulting different rankings were used to select a different set of sites than would have been selected by the FPC group alone. Third, it encouraged the FPC group to modify its criteria set to match the EAG set and to modify its weights in the direction of the EAG weights. Those modified criteria and weights were used in the FPC ranking and selection of potential sites for Phase III.

The EAG role in the Phase IV ranking and Phase V selection closely paralleled its role in Phase II. The EAG added two criteria in Phase IV: "Potential for Broader Purpose" and "Regional Urbanization." It recognized that at some sites in particular, the power plant could fulfill "broader purposes," such as: the use of damaged land that would otherwise remain a blight; the basis for conservation easements or other forms of land preservation; the use of waste for fuel or sewage for cooling; and heat cogeneration. A three-point scale was developed for rating potential for broader purpose. The EAG also recognized that a power plant could be a large burden on the infrastructure of a remote area. Therefore, it suggested a criterion favoring sites with development near enough to provide infrastructure support. The measure selected was size of population within 9 to 25 miles. Phases III and IV reduced the number of sites from 61 to 21 to 6.

In Phase V, each group confirmed the ranking of the six candidate sites and then selected the preferred and two alternate sites on the basis ofthat ranking. Rankings in Phases IV and V were done with 11 criteria that were elicited, along with their weights, jointly from the FPC group and the EAG.

Additional Features of the Case

The iterative process used in this instance had several benefits. First, it allowed an efficient targeting and use of data collection and analysis resources. The entire study area was first examined at a coarse scale of data, and more promising potential sites were examined at a finer level of detail. Money and time were saved by not collecting detailed data about unpromising areas. Second, iteration between the FPC group and the EAG allowed the former to revise its own criteria and weights. This process moved the two groups' judgments closer together. Third, the process built trust among the members of the EAG, who initially suspected that their inputs would not make a difference. The use of EAG input to revise the map of areas to be considered for further study convinced EAG members of the intention of FPC to actually use their inputs in the selection of the sites.

The value elicitation procedure resulted in a useful value tree graphic that looks like an organization chart. The graphic helped the participants by clearly relating the overall mission of the project to the criteria and the specific indicators used to assess the adequacy of possible sites. The graphic and the process that generated it made clear multiple perspectives. Building a weighted list of concrete criteria and indicators focused discussion, thereby discouraging posturing and vague rhetoric. People directed their concerns toward the goal of improving the collective value tree.

In the FPC case, multiattribute utility analysis techniques helped incorporate multiple concerns into the site-selection decision and resulted in a transparent decision-making process that has acquired a high degree of public acceptance. The licensing for the preferred site is proceeding in a relatively smooth way. Using this analytic approach does not, however, guarantee that all participants' ideas are meaningfully included in an analytic-deliberative process or that the resulting rankings are meaningful (Brody and Rosen, 1994). Neither does this approach, even when used sensitively, guarantee noncontroversial siting of contentious facilities when it is becoming increasingly difficult to license any large, locally unpopular facility. Nevertheless, it is encouraging that people from diverse perspectives could reach agreement on the siting of a 3,000 megawatt coal-fired power plant. Another encouraging sign is that two of the members of the EAG—in fact, two who would in other circumstances be expected to oppose actions by FPC—have gone on public record on their own initiative in support of the siting process.

THE CALIFORNIA COMPARATIVE RISK PROJECT

When California began its comparative risk project (CCRP), four states had already completed similar studies and 10 others were under way. Comparative risk projects work from the assumption that policy priorities of environmental problems should be determined, at least in part, by the magnitude of the risk each problem presents (U.S. Environmental Protection Agency, 1987). While the CCRP workplan (California Environmental Protection Agency, 1992:i) emphasized ranking risks using "good scientific minds together to help establish the 'best science,'" it also realized that deliberation among scientists would not be a sufficient basis to set policy priorities. The California project paid explicit attention to the need for broader participation in its initial process design (California Environmental Protection Agency, 1992:2):

> The responsibility for protecting California's environment applies not only to government, but also depends on the involvement of individuals with academic, industrial, business, activist, residential, and political interests within the State.... We will conduct the comparative risk project in such a way as to allow for all opinions to be accounted for, since the project is dedicated to expanding the public's ability to make important decisions about the fate of their environment.

Direct citizen participation did not play as large a role as this statement might suggest. Still, the CCRP was significant because it recognized the need for broadly participatory deliberation and for a broad agenda for risk analysis—it presumed that social, economic, and equity concerns would have to be included in any risk ranking scheme. This was evident in both the way the problem was formulated and the fact that the CCRP was willing to modify its process design as a result of deliberation about whether it would produce the needed information.

The initial process design developed by the California EPA as a result of its diagnosis of the situation called for three technical committees—on human health, ecological health, and social welfare—to work independently to rank risks in their categories, while three other committees would consider management options and the legal and economic constraints associated with making those choices (California Environmental Protection Agency, 1992). In a second phase, the risk rankings of the three committees were to be integrated during a two-day statewide symposium. This design seemed to have analysis preceding deliberation and to confine the participation of nontechnical people to the decision-making phase of the project.

Soon after the project began, this process design was challenged on the grounds that the technical committees could not be expected to produce purely objective risk rankings. The technical analysis would be permeated, some said, with policy considerations at various levels. One argument was that the use of population-risk estimates left risk managers blind to the inequitable distribution of risk among subpopulations. This concern led to the creation of an environmental justice committee to address such issues. The original process design was also altered to include more feedback and interaction among the technical committees and to include critics of a conventional risk analysis approach.

The new process design comprised two "components" operating in parallel with significant cross-fertilization. One involved the original three technical committees, using their knowledge of quantitative risk assessments in order to rate the impacts of various environmental conditions on the broad areas of human health, environment, and social welfare. The other comprised three committees that supplemented the technical risk assessments with social and economic concerns. One of these committees, the environmental justice committee, raised additional concerns (e.g., social equity) for inclusion in the risk rankings and also additional options (pollution prevention) for consideration (California Environmental Protection Agency, 1994). In doing these things, the committee called into question the initial problem formulation as too narrow. Its perspective found its way into the criteria finally used to rank the social welfare and human health impacts of environmental stressors (equity was one of the criteria adopted, although pollution prevention was not), and into the CCRP's final report in the form of committee recommendations and a "critique of the risk-ranking model."

Each technical committee reached a surprising level of agreement on rankings considering the great diversity of backgrounds among each committee's members. Nonetheless, the CCRP generated considerable controversy, particularly over some of the more subjective social welfare outcomes, such as "peace of mind," that it endorsed as important (Stone, 1994). The CCRP's statewide Community Advisory Committee decided not to integrate all three rankings into a common list because of time constraints, concerns over technical issues involved in reducing the three very different rankings into a single ranking, and discomfort about including in any "technical" ranking what some characterized as "subjective" and "nonscientific" social welfare rankings. Some trade organizations went directly to the press and the governor's office with the concern that "California has come up with a new and controversial method of evaluating environmental risk that downplays the traditional role of science and takes into account people's values, opinion, fears, and anxieties" (Clifford, 1994:A1). In election year politics, the state government distanced itself from the CCRP report: although it released the report, it did not publicize its results or apparently use its rankings in setting priorities.

Despite this immediate outcome in state politics and the fact that the project did not fulfill all of its own objectives, the project is instructive because it brought together conventional forms of risk analysis with analysis and deliberation about various social, economic, equity, and other concerns. It demonstrates the importance of iteration in process design, particularly the use of deliberation to revisit the provisional problem formulation and the process design that emerged from the diagnostic phase, and the links between process design and problem formulation. The new process did not reduce controversy, but it did make explicit some of the different views of the nature of environmental problems underlying environmental policy conflicts in the state.

The longer-term effects of the project remain to be seen. One possibility is that the project's methods and findings may be reconsidered at a later date. Another, less sanguine, possibility is that events may have made it more difficult to organize future broadly based risk deliberations in California because many participants who worked hard on the project may have been alienated. Some saw concerns they judged to be legitimate and essential successfully painted as unscientific and therefore unworthy of consideration by others (some of whom also participated), working outside the process. Some volunteered considerable time and effort to technical analyses and deliberations and, acting on good faith, agreed with others (with whom they publicly disagreed in other forums) on rankings, only to see their hard work set aside by the political process.

PLANNING FUTURE LAND USES AT HANFORD, WASHINGTON

Hanford is a 560-square-mile site on the Columbia River that was used for decades by the U.S. Department of Energy (DOE) and its predecessors for the production of nuclear materials for national defense. Air, water, and soil at the site have been contaminated with radioactive materials. Since the mission of the site ended in the 1980s, attention has turned toward restoring its environment and preparing the location for future uses. As part of this project, the DOE is required to complete an environmental impact statement (EIS) that would determine potential impacts associated with the cleanup and restoration projects. DOE decided to seek participation by representatives of a wide range of governments (federal, state, tribal, local) and various other organizations and interest groups in planning the EIS and identifying alternative scenarios for future use of the site.

Customarily, an EIS begins with a scoping process to identify all environmental issues that are likely to be significant in the assessment. Scoping is meant to make the impact analysis more efficient by focusing attention on realistic issues and concerns. It involves defining assessment objectives and is similar to problem formulation as described in this report. Risk assessments are likely to be part of any EIS concerning a hazardous waste site, and the scoping process is a way of ensuring that the policy options and possible harms addressed in the assessments are consistent with the needs of the decision makers and affected parties.

DOE, in collaboration with EPA, the Oregon and Washington state governments, and county and tribal governments of the region composed a list of potential participants for the Hanford Future Site Use Working Group, including both public officials and interested and affected parties, to advise on the EIS (Hanford Future Sites Working Group, 1992). One of the most noteworthy features of this effort was that the process design for the working group was not imposed in advance, but was created only after conducting interviews with the prospective members. Each candidate was asked what would make the process successful, and each was also asked to nominate other potential participants. By obtaining this feedback at the start, the organizers bettered the chances that the eventual EIS would serve the needs of the interested and affected populations. Three things emerged from the interviews as important: the process should consider a wide variety of viewpoints, it should provide a common base of information about the site, and the decision-making agencies should commit to using the products of the process in their decision making.

With the assistance of a professional facilitation team, the working group drafted its own charter, specifying its purpose and the scope of its work. The main task was to identify alternative scenarios for cleanup and future site use, based on the participants' visions and influencing factors. The group agreed to focus on how cleanup and future site uses would be connected to each other—essentially specifying some outcomes to be subjected to analysis in the EIS. The working group also decided not to seek consensus on a preference for future site uses, but to emphasize commonalities among the options generated. It also decided to make all decisions by consensus.

A significant feature of the Hanford Future Site Uses Working Group was that the participants identified the information they needed. Since each participant represented a specific constituency, the information needs were linked with the participants' objectives. After reviewing existing information and visiting the site, the group composed a list of needed educational seminars and information. This resulted in 34 presentations by expert teams at various meetings of the working group. A huge amount of information is available on the history, use, and present condition of the Hanford site. While the working group noted important gaps in data and potential uncertainties, it did not have the resources or time to initiate any specific studies. However, it did identify needs for future data collection and analysis.

The working group met five times over a 6-month period, and participants were expected to consult with their constituencies in the time between meetings to get wider input. The group issued draft recommendations with consensual support and brought them to the public for review and comment at eight "open houses" in various locations in Washington and Oregon. The draft report was revised to reflect the comments received.

The working group made nine major recommendations in addition to specifying general future use options and cleanup scenarios for different areas of the Hanford site. Several of the general recommendations had to do with possible outcomes that the working group asserted should be strongly considered in any impact assessment: treaty rights of Native American tribes (especially access to religious sites); stopping and preventing contamination of the Columbia River and cleaning up groundwater; access to contaminated areas; public awareness of waste shipments; and jobs and regional economic impacts. The working group did not offer a complete list of possibly significant outcomes, nor did it rank their importance.

The bulk of the working group's recommendations concerned: (a) defining eight general future use options for each subarea of the Hanford site, (b) listing some specific and important possible outcomes relevant to that area (e.g., effects on cultural sites, wildlife, and industry), and (c) defining the cleanup scenarios necessary to assure the safety of each future use. Cleanup levels were delineated according to access restrictions, which would be based on health risk assessments.

This case illustrates how a risk decision-making process can use techniques for deliberation and public review in combination with analysis to arrive at a widely acceptable formulation of the problem and to design a process that is likely to characterize risks in a way that meets the needs of the interested and affected parties and public officials. DOE is seriously considering the working group's recommendations. Before preparing the draft EIS, DOE staff prepared an implementation plan that assimilated the recommendations of the working group. In a truly iterative process, the DOE staff reconvened the working group 1 year after its dissolution to confirm that its recommendations had been properly interpreted by DOE in the EIS implementation plan. The entire working group recommendation document was included in the draft EIS scoping chapter, as the working group had requested. DOE has found the process valuable enough to require similar future use working groups at all of its cleanup sites. DOE's reliance on local public involvement, however, remains controversial (see, e.g., Blush and Heitman, 1995), and the ultimate effect of the process on cleanup at Hanford remains to be seen. Because the underlying issues remain intensely controversial, a policy consensus may remain elusive even if the EIS addresses all the parties' questions.

Common Approaches to Deliberation and Public Participation

The literature on public participation identifies numerous techniques that have been used in decision making about risks. For example, a recent report focusing on remediation of contaminated sites listed 13 major approaches and several dozen more limited approaches (English et al., 1993; for another typology see Renn, Webler, and Wiedemann, 1995). Some of the techniques are clearly deliberative in nature, while others collect public input without involving interested and affected parties in deliberations among themselves or with public officials. It is important to note that there is no generally agreed typology. Some of the techniques contain common elements, and practices that fall under the same general heading can vary greatly in implementation. The following listing identifies several major types of techniques of deliberation and public participation. It includes a range of methods that can be used, notes important considerations related to each, and identifies supplementary sources of information about the methods.

Public Hearings

Public hearings often are held during development of legislation and sometimes during development of implementing regulations. They are also mandated by a variety of laws (e.g., Superfund). "Public meetings" are similar to public hearings in format, but they are not mandated by law.

Strengths: Public hearings meet legal requirements for participation, are relatively easy to convene, have the potential to reach many interested and affected parties, and may enable these parties uninterrupted time to present their views.

Concerns: Public hearings are often pro forma and occur too late for input to be meaningful to an organization. They invite posturing and make it difficult to hold meaningful discussions. Public hearings and meetings are frequently used by organizations to "decide, announce, and defend" their risk characterizations, a mode of iteration that does not lend itself to open-minded exchange.

In some instances organizations have overcome these problems by drafting agendas for public meetings in consultation with interested and affected parties, by using neutral facilitators to encourage dialogue, and by using public hearings as merely one element of a larger deliberative process. Public hearings are best for presenting alternative views, information, and concerns; they are less useful for dealing with power imbalances, creating trust, or promoting dialogue. There is rarely a decision or consensus at the end of such meetings, allowing agencies great latitude to ignore any comments, which can create further mistrust.

For further information, see Heberlein (1976), Mazmanian and Nienaber (1979), Checkoway (1981), Rosener (1982), U.S. Environmental Protection Agency (1983), Hadden (1989), and Webler and Renn (1995).

Citizen Advisory Committees and Task Forces

Individuals are appointed to advisory committees and task forces by an organization to consider an issue or issues. Task forces tend to be created to consider a specific issue in depth, with the goal of providing recommendations, after which the task force is disbanded. Citizen advisory committees have indefinite life spans. Otherwise, the two kinds of groups are about the same.

Strengths: Continuously meeting groups can build a common base of information, can discuss complex issues, can build relationships, and can promote better mutual understanding of the concerns of the sponsoring agencies and the group members.

Concerns: Who appoints the advisory committee or task force? Do participants represent opinions of organized interest groups or the larger population? Who determines the agenda? Do all the interested and affected parties have time and expertise to participate fully and effectively? Will the agency take committee recommendations seriously?

Theoretically, citizen advisory committees and task forces can address many of the deficiencies of public hearings because of their continuity. They usually use consensus-based decision rules, which encourages participants to find common ground or develop novel solutions. If a citizens advisory committee or task force is provided with the information that members want, access to appropriate agency personnel, or an independent technical adviser as needed, it may be extremely productive. However, unempowered groups (e.g., low income populations) and those with views that diverge too far from those of the organization are often excluded, undermining the legitimacy and credibility of the approach.

Managers may consider potential responses of an advisory committee or task force in advance of making routine decisions. However, a manager may also ignore a functional task force or even disband it before it makes recommendations.

For further information, see Pierce and Doerksen (1976), Creighton (1980), Anderson (1986), Lynn (1987a), Houghton (1988), Ross and Associates (1991), Lynn and Busenberg (1995), and Lynn and Kartez (1995).

Alternative Dispute Resolution

Various means of informal or alternative dispute resolution have become increasingly popular for achieving consensus among government agencies and interested and affected parties (e.g., Crowfoot and Wollendeck, 1990). Alternative dispute resolution may take the form of mediation, which seeks to bring agencies and the interested and affected parties together to reach consensus through a facilitated process. Mediations have occurred on a range of risk-related issues and have involved local, state, and federal agencies. Mediation usually requires the involvement of the spectrum of interested and affected parties for any agreement to be implemented without determined opposition. Unless the parties feel they have affected the decision, it is not likely to be satisfactory to them. For research on mediation, see Kressel and Pruitt (1985), Susskind and Ozawa (1985), Bingham (1986), Cormick (1987), Dryzek and Hunter (1987), Susskind and Cruikshank (1987), Stephenson and Pops (1989), and Baughman (1995).

Policy dialogues are a form of alternative dispute resolution that can involve agencies and interested and affected parties in discussion of controversial issues. These dialogues seek to develop better understanding of differences among parties and may not be geared to the development of formal agreements. Instead, the dialogue may build common ground that may then serve as a foundation for policy development, regulation, or further interactions among the parties. For further information, see Bingham (1986) and Gray (1989).

In contrast, the goal of negotiated settlements is to reach formal agreements. For example, regulatory negotiation can take place during the development of legislation or implementing regulations. In regulatory negotiation, representatives of interested and affected parties work consensually with government regulatory bodies to draft a proposed rule. Once completed, the rule is formally proposed by the agency (if it is a federal agency, in the *Federal Register*) and subjected to the normal process of public comment and review. The U.S. Environmental Protection Agency (EPA) has frequently used regulatory negotiation to draft complex and highly technical rules, especially when there is a clear need for a rule but insufficient data to support the customary EPA rule-making process. The purpose of regulatory negotiation is to reduce legal challenges to new rules by involving would-be adversaries directly in the rule-making process and by producing a draft rule that meets legal requirements and is acceptable to a wide array of interested and affected parties.

Participants are selected in a regulatory negotiation because they represent major and important interest groups and sometimes also because they have relevant scientific expertise. Expertise is necessary because regulations are highly technical, and negotiators need to be well versed in the scientific literature, theories, methods, and data. At the same time, participants must negotiate on issues in which there are gaps in theory or in data or on issues that involve tradeoffs that cannot be resolved with scientific means, such as between different kinds of adverse outcomes. Thus, participants also need skills of persuasion, argumentation, and negotiation.

For further information on regulatory negotiation, see Harter (1982), Susskind and McMahon (1985), Wald (1985), Perritt (1986), Funk (1987), Fiorino (1988, 1991, 1995), Rushefsky (1991), Kelman (1992), and Hadden (1995).

Strengths: Alternative dispute resolution can deal with complex issues, strongly held beliefs, polarized opinion, conflicting values, and technical concerns.

Concerns: Will parties accept alternative dispute resolution rather than litigation, direct action, or delay? Are the right parties at the table? Is the playing field level? Is there sufficient commitment to the process?

Policy makers, who may have been part of a mediation or negotiation, can act more easily afterwards, and litigation is reported to be less likely (Susskind and Cruikshank, 1987). However, alternative dispute resolution may exacerbate power imbalances and increase conflict if it obscures the needs of those who were not part of the mediation or negotiation. It may also be successful from a process perspective, for example, by reaching an agreement, but ignore critical analytic issues, possibly resulting in an agreement on an unwise option.

Citizens' Juries and Citizens' Panels

Citizens' juries rely on a randomly selected pool of citizens to evaluate policy alternatives. Typically, citizens juries are asked to express a preference among three or four policy options. Staff from, or selected by, an organization set the charge and usually impose the principle of majority vote as the means used to resolve conflicts. Majority vote does not guarantee the integrity of minority interest positions in the discourse. An oversight committee may be created to set the agenda, the rules for discourse, and oversee the process, but the membership of the oversight committee is limited to those selected through random sampling procedures.

Strengths: Citizens juries represent interested and affected parties' knowledge, positions, and perspectives, largely by virtue of the random sample. Competence in citizens' juries is encouraged by face-to-face mediated discussions in which everyone has an equal opportunity to make statements, challenge others' statements, and express opinions. Citizens' juries consider and often welcome and respect anecdotal evidence; they may also use peer review of technical information packages.

Concerns: The citizens' jury approach does not necessarily promote critical inquiry into the factual issues or use a systematic method to reach the best possible understandings about facts and states of affairs. Crosby (1995) recognizes a need to change this, but notes the issue of cost.

Participants in citizens' juries are free to ask their own questions to "witnesses" who come before the jury, thus collecting knowledge they deem relevant and verifying understandings. The format of the discourse also promotes normative inquiry and debate among the jurists. However, systematic methods to arrive at shared preferences about the available normative choices are not normally used. Citizens' juries do not usually structure values in any formal way, make impact assessment profiles on the value dimensions, or evaluate the consistency of the outcome with established norms and laws. These more structured techniques could, however, be combined with the citizen jury approach (e.g., Renn et al., 1991, 1993); for more information, see Crosby, Kelly, and Shaeffer (1986), Crosby (1995), Armour (1995).

Citizens' panels are similar to citizens' juries, except that they tend to be given a freer hand in identifying the options to be considered, rather than deliberating on a given set of options. This approach has been developed in Europe (Dienel, 1989; Renn et al., 1993; Dienel and Renn, 1995), but some applications in the United States adopt its essential elements (e.g., Stewart, Dennis, and Ely, 1984; Kathlene and Martin, 1991). The success of both citizens' panels and citizens' juries relies heavily on a clear mandate from the legal decision-making organization to seriously consider implementing the recommendations of the panel or jury; the involvement of randomly selected citizens; and legitimacy of the randomly selected citizens in the eyes of stakeholders and other citizens. It may be suspected that citizens in juries would be too shy to ask critical questions of experts, regulatory officials, and interest group representatives. After all, these are the people who usually make the decisions. But experience shows that citizens can fulfill very well the task of "value consultants" (Crosby et al, 1986; Kathlene and Martin, 1991). They read background material, seek out needed information, and consider the subject matter very seriously.

For additional information, see Crosby, Kelly, and Schaeffer (1986), Jefferson Center (1988), Crosby (1995), and Armour (1995).

Surveys

Randomly selected citizens complete one or more surveys about an issue (Milbrath, 1981).

Strengths: The method is dispassionate, it can elicit opinions from many people, and no specialized knowledge is needed to participate.
Concerns: Are responses shaped by questions? Do questions oversimplify the choices? Do agencies use surveys to avoid dealing with people, values, etc? Who interprets the responses? Surveys elicit public comment, but not deliberation among the participants.

Surveys can help organizations understand perceptions, knowledge, and demographic variations in people's views, and they can provide public input about options. However, the organization shapes the questions, rarely publicizes the results, and makes decisions without interacting further with those surveyed. Thus, the approach may exacerbate distrust. Depending on sampling, surveys can obscure or highlight needs of special populations. Survey results may influence decisions (e.g., avoiding a policy that will anger many, justifying a policy that suits the "silent majority," or crafting alternative options based on survey input). However, as one study showed, officials can easily ignore the results (Milbrath, 1981).

Focus Groups

Focus groups, or more precisely, focused group discussions, are used extensively by marketing firms and increasingly by government agencies and other organizations. They typically include 6-10 members selected more or less at random. The groups are usually homogenous, consisting of members recruited from targeted populations to deal with a specific, usually nontechnical subject (e.g., homeowners' view about radon). The groups, which are led by a facilitator who asks set questions, discuss but do not decide or recommend. The discussions are used to understand outside views, appreciate potential reactions to policies, or pretest written materials. Focus groups are used increasingly, in part because they are low-cost, no-risk means of eliciting opinion. The results of the discussions may not be made public, and the organization retains all decision-making power. In this respect focus groups share many of the strengths, limitations, and applications of surveys.

Interactive Technology-based Approaches

A number of technology-assisted methods of deliberation have been developed; they have not been widely used in government, with the possible exception of the Army Corps of Engineers. One example is computer-generated models, which can simulate the results of policy options. Participants can manipulate models to help generate or consider policy alternatives. Another example is computer-assisted meetings, in which each participant has a device to send input to a central computer. A moderator can then ask questions throughout the discussion, and the participants can vote. The votes can be immediately tallied and displayed.

Combinations of Deliberative Methods

Public participation programs may use different deliberative methods at different stages of the process or with different audiences. For example, to develop water quality regulations, the New Jersey Department of Environmental Protection promoted a multifaceted deliberative process, solicited input on process as well as substance, developed a task force of interest groups, held informal meetings with various stakeholder groups, distributed a survey to 500 people for consideration of various thorny issues, and ultimately held mandated public hearings on the proposed regulations (Chess, 1989). The regulations were received positively.

Renn and his colleagues (Renn et al., 1993) have developed a three-step model that has reportedly worked well in Europe. This approach involves interest groups in generating a "value-tree analysis" that identifies their concerns and allows each group to weigh those concerns. Experts then participate in a modified Delphi process in which they make judgments about how each of the options will affect the outcomes of concern to the interest groups. Finally, a citizen panel (usually selected at random) develops a report and a set of recommendations for action, based on a deliberative process in which it considers the results of the Delphi process, presentations by experts, further fact finding, and the views of the panel members.

Biographical Sketches

HARVEY V. FINEBERG (*chair*) is dean of the Harvard School of Public Health. He has served on the Public Health Council of Massachusetts, as chair of the Health Care Technology Study Section of the National Center for Health Services Research, as president of the Society for Medical Decision Making, as a consultant to the World Health Organization, and as member or chair of a number of Institute of Medicine panels dealing with topics of health policy. His research has focused on health policy, including the process of policy development and implementation, assessment of medical technology, and dissemination of medical innovations. He is coauthor of *Clinical Decision Analysis* (with Milton C. Weinstein and others) and *The Epidemic That Never Was* (with Richard E. Neustadt), an analysis of the controversial federal immunization program against swine flu in 1976. In 1988 he received the Joseph W. Mountin Prize from the Centers for Disease Control and the Wade Hampton Frost Prize from the Epidemiology Section of the American Public Health Association. He received A.B., M.P.P, M.D., and Ph.D. degrees from Harvard University.

JOHN AHEARNE is director of the Sigma Xi Center, The Scientific Research Society, and lecturer in public policy at Duke University. He has served in government as Deputy Assistant Secretary of Defense, Deputy Assistant Secretary of Energy, Chairman of the U.S. Nuclear Regulatory Commission, and in advisory positions to the U.S. Departments of Energy and Defense and the U.S. General Accounting Office. He has also served as vice president of Resources for the Future and as chair and member of several National Research Council committees and panels. He is a member of the National Academy of Engineering and a fellow of the American Physical Society. He received a B.S. in engineering physics, an M.S. from Cornell University, and M.A. and Ph.D. degrees in physics from Princeton University.

THOMAS A. BURKE is an associate professor of Health Policy and Management and codirector of the Risk Sciences and Public Policy Institute at the School of Hygiene and Public Health of the Johns Hopkins University. His research interests include environmental epidemiology, the evaluation of population exposures to environmental pollutants, assessment of environmental risks, and the application of epidemiology and health risk assessment to public policy. Prior to his appointment at Johns Hopkins, he was Deputy Commissioner of Health for the state of New Jersey. He has served as a member of the Council of the Society for Risk Analysis, an adviser to the Office of Technology Assessment on risk assessment of chemical carcinogens and managing nuclear materials, and a member of the National Research Council Committee on Remediation of Buried and Tank Wastes. He received a B.S. degree from Saint Peter's College, an M.P.H. from the University of Texas, and a Ph.D. in epidemiology from the University of Pennsylvania.

CARON CHESS is director of the Center for Environmental Communication at Rutgers University. Her research interests include methods to evaluate public participation in environmental policy decisions and exploration of the internal organizational factors that influence risk communication and public participation efforts. She has coauthored a variety of handbooks and related materials for government agencies, including *Improving Dialogue with Communities: A Short Guide for Government Risk Communication*, which is widely used in the United States and has been translated into two languages for use abroad. Before moving to academia, she coordinated programs for government and nonprofit organizations, including playing a central role in the campaign for the country's first right-to-know law. She received an M.S. degree in environmental communications from the University of Michigan.

BRENDA S. DAVIS is vice president, government operations, and a member of the Management Board of Johnson & Johnson Health Care Systems, Inc. In that position she is responsible for government sales, state government affairs, reimbursement services, and pharmaceutical rebate management for the domestic health care businesses. Previously, she was a visiting fellow at Princeton University, served in the cabinet of Governor Thomas H. Kean of New Jersey, and was a senior staff member of the

Committee on the Budget of the U.S. Senate. She received a Ph.D. degree in ecology from the University of California at Berkeley.

PETER L. DEFUR is an affiliate associate professor in the Center for Environmental Studies at Virginia Commonwealth University and an adjunct senior scientist with the Environmental Defense Fund in Washington, D.C. Previously, he held faculty positions at Southeastern Louisiana University and George Mason University in Virginia. His interests and research have covered adaptations of aquatic animals, especially in coastal waters; the application of scientific information and process to environmental policy and regulation; and, most recently, approaches to ecological risk assessment and chemical threats to human and environmental health, especially from chemicals that interfere with hormonal systems in wildlife and humans. He is on the editorial board of the *Journal of Experimental Zoology*, cochair of the steering committee of the Science and Environmental Health Network, and a board member of the Coalition to Restore Coastal Louisiana. He received a Ph.D. degree in biology from the University of Calgary.

JEFFREY HARRIS is a primary-care internist at Massachusetts General Hospital and a professor of economics at the Massachusetts Institute of Technology. He was a member of the first Institute of Medicine committee on AIDS, the Committee on National Strategy Toward AIDS. He has served on the Institute of Medicine's committee on strategies to reduce low birthweight and on the National Research Council's committee on diesel emissions. He has advised numerous public and private agencies on issues of risk management, health economics, and public policy and testified before the House Committee on Ways and Means and the Massachusetts legislature. He wrote the seminal chapter in the 1989 Surgeon General's Report, which estimated that smoking caused nearly 400,000 deaths annually, and he is the author of *Deadly Choices: Coping with Health Risks in Everyday Life*. He received an A.B. degree from Harvard University and M.D. and Ph.D. degrees from the University of Pennsylvania.

MARK A. HARWELL is director of the Center for Marine and Environmental Analyses, Rosenstiel School of Marine and Atmospheric Science, University of Miami. Previously, he was associate director of the Cornell University Ecosystems Research Center. He is an ecosystems ecologist, specializing in ecosystem modeling and developing methods for ecological risk assessment and ecosystem management and applying ecological principles to real-world environmental problems. Dr. Harwell has directed major national and international research programs on high-level nuclear waste disposal, global environmental consequences of nuclear war, ecological and agricultural consequences of global climate change, methodologies for ecological risk assessment and environmental decision-making and human/environmental issues of ecosystem management and ecological sustainability. He is a member of the Science Advisory Board of the U.S. Environmental Protection Agency and chair of the Human-Dominated Systems Directorate of the U.S. Man and the Biosphere Program (US MAB). He received a Ph.D. degree in ecosystems ecology from Emory University.

SHEILA JASANOFF is professor of science policy and law and chair of the Department of Science and Technology Studies at Cornell University. Her research explores the relationship between science, politics, and the legal system, with specific attention to risk management, environmental regulation, and comparative and national science and technology policy. She has held visiting appointments at Yale University, Harvard University, Wolfson College (Oxford), and Boston University School of Law. She is a recipient of the distinguished achievement award of the Society for Risk Analysis and an editorial adviser to *Social Studies of Science, Science, Technology, and Human Values, Science and Engineering Ethics*, and *Environmental Science & Technology*. She received a Ph.D. degree from Harvard University and a J.D. degree from Harvard Law School.

JAMES C. LAMB IV is vice president, scientific and technical services, at the environmental consulting firm of Jellinek, Schwartz & Connally, Inc. He advises clients on scientific issues and regulatory and science policies, specializing in general toxicology, carcinogenesis, reproductive and developmental toxicology, risk assessment, and regulatory policy. Previously, he was Special Assistant to the Assistant Administrator for Pesticides and Toxic Substances at the U.S. Environmental Protection Agency and head of the fertility and reproduction group of EPA's National Toxicology Program. He is a lawyer and a board-certified toxicologist and past president of the American Board of Toxicology. He received a Ph.D. degree in pathology from the University of North Carolina at Chapel Hill and a J.D. degree from the North Carolina Central University School of Law.

D. WARNER NORTH is a senior vice president of Decision Focus Incorporated, a consulting firm in Mountain View, California, specializing in management science and quantitative risk analysis, and consulting professor in the Department of Engineering-Economic Systems at Stanford University. He has carried out applications of decision analysis and risk on management of toxic substances in the environment, quarantine policy for the exploration of Mars, wildland fire protection, weather modification, nuclear waste disposal, and environmental impacts from energy technologies. He serves as a member and consultant to committees of the EPA Science Advisory Board, and has served as a presidentially appointed member of the Nuclear Waste Technical Review Board, and as a member of the Scientific Advisory Panel on toxic substances under Proposition 65 for the governor of California. He is a past president of the Society for Risk Analysis. He has been a committee member for many previous National Research Council reports dealing with risk, is currently a member of the Board on Radioactive Waste Management, and serves as chair of the Transportation Research Board's review of federal estimates of the relationship of vehicle weight to fatality and injury risk. He received M.S. degrees in physics and mathematics and a Ph.D. degree in operations research from Stanford University.

KRISTIN SHRADER-FRECHETTE is distinguished research professor at the University of South Florida in the Program in Environmental Sciences and Policy and in the Department of Philosophy. She previously held professorships at the University of Florida and the University of California and has held postdoctoral fellowships from the National Science Foundation in ecology, economics, and hydrogeology. She specializes in analysis of ecological methods, environmental ethics and policy, and quantitative risk assessment, including ecological risk assessment. She is a member of the National Research Council Board on Environmental Studies and Toxicology and president-elect of the Risk Assessment and Policy Association. Author of many articles in biology, risk assessment, and philosophy journals, her three most recent books are *Method in Ecology, Burying Uncertainty: Risk and the Case Against Geological Disposal of Nuclear Waste* and *Ethics of Scientific Research*. Shrader-Frechette is also associate editor of *BioScience* and editor-in-chief of the Oxford University Press Series "Environmental Ethics and Science Policy." She received undergraduate degrees in mathematics and physics and a Ph.D. degree in philosophy of science from Notre Dame.

PAUL SLOVIC is president of Decision Research and a professor of psychology at the University of Oregon. He studies human judgment, decision making, and risk analysis. He and his colleagues worldwide have developed methods to describe risk perceptions and measure their effects on individuals, industry, and society. They created a taxonomic system that enables one to understand and predict perceived risk, attitudes toward regulation, and impacts resulting from accidents or failures. He publishes extensively and serves as a consultant to many companies and government agencies. He is past president of the Society for Risk Analysis and in 1991 received its Distinguished Contribution Award. He also serves on the Board of Directors for the National Council on Radiation Protection and Measurements. In 1993 he received the Distinguished Scientific Contribution Award from the American Psychological Association, and in 1995 he received the Outstanding Contribution to Science Award from the Oregon Academy of Science. He received a B.A. degree from Stanford University and M.A. and Ph.D. degrees from the University of Michigan.

MITCHELL J. SMALL is a professor of civil and environmental engineering and of engineering and public policy at Carnegie Mellon University; he serves as associate department head for graduate education in engineering and public policy. His research interests include mathematical modeling of environmental quality and exposure, statistical analysis of monitoring data, and methods for uncertainty and decision analysis. He has served as a member of the Science Advisory Board of the U. S. Environmental Protection Agency. He is currently an associate editor for policy analysis for *Environmental Science & Technology*. He received a Ph.D. degree in environmental and water resources engineering from the University of Michigan.

PAUL C. STERN is study director of the Committee on Risk Characterization at the National Research Council, where he also serves as study director for the Committee on the Human Dimensions of Global Change and the Committee on International Conflict Resolution. He is also a research professor of sociology at George Mason University. He previously staffed several National Research Council committees, including the one that produced *Improving Risk Communication*. His major research interest is in the human dimensions of environmental problems. He is coauthor of a textbook, *Environmental Problems and Human Behavior* (with Gerald T. Gardner) and has published on behavioral aspects of residential energy conservation, attitudes and values as they affect environmentally significant behavior, the psychological dimensions of global environmental change, social science research methods, international conflict, and nationalism. He holds a B.A. degree from Amherst College and M.A. and Ph.D. degrees in psychology from Clark University.

ELAINE VAUGHAN is associate professor of psychology in the School of Social Ecology at the University of California, Irvine. Previously, she served as a research psychologist in the Division of Adolescent Medicine at the University of California, San Francisco. Her research focuses on interactions among cognitive, sociocultural, economic, and environmental factors that affect the risk judgments and behaviors of individuals and implications of group differences in responses for social policies. She has received an award for excellence in interdisciplinary research from the School of Social Ecology and has served as a committee member for California's Comparative Risk Project (1992-1994). She received a Ph.D. degree in social psychology from Stanford University.

JAMES D. WILSON is senior fellow and leader of the risk analysis program in the Center for Risk Management at Resources for the Future. Previously, he had a long career with the Monsanto Company in research, research management, and health and environmental policy. His research has focused on structure-activity relationships, including environmental chemistry broadly, dioxin and related chemicals, the relationship of chemical structure to physical and physiological properties, the use of science in decision making, and the influence of organizational structure on decision making. He was president of the Society for Risk Analysis in 1993 and was named a fellow of the Society in that year. He received a Ph.D. degree in organic chemistry from the University of Washington.

LAUREN ZEISE is chief of reproductive and cancer hazard assessment at the California Environmental Protection Agency. She has also worked at the California Department of Health Services and the California Public Health Foundation. Her research has focused on cancer risk assessment, particularly models of exposure. She received a B.S. degree from Loyola University and S.M. and Ph.D. degrees from Harvard University.

Glossary

Affected parties.   People, groups, or organizations that may experience benefit or harm as a result of a hazard, or of the process leading to *risk characterization*, or of a decision about *risk*. They need not be aware of the possible harm to be considered affected.

Analysis.   The systematic application of specific theories and methods, including those from natural science, social science, engineering, decision science, logic, mathematics, and law, for the purpose of collecting and interpreting data and drawing conclusions about phenomena. It may be qualitative or quantitative. Its competence is typically judged by criteria developed within the fields of expertise from which the theories and methods come.

Broadly based deliberation.   *Deliberation* in which participation from across the spectrum of *interested and affected parties*, by policy makers, and by specialists in *risk analysis* is sufficiently diverse to ensure that the important, decision-relevant knowledge enters the process, that the important perspectives are considered, and that the parties' legitimate concerns about the inclusiveness and openness of the process are addressed. Such deliberation involves the participation or at least the representation of the relevant range of interests and values as well as of scientific and technical expertise.

NOTE: When definitions refer to other defined terms, the latter appear in italics.

Deliberation .   Any process for communication and for raising and collectively considering issues. In the process leading to *risk characterization*, deliberation may involve various combinations of scientific and technical specialists, public officials, and interested and affected parties, and may be formalized (as in mediation) or occur in informal settings. It may be used both to increase understanding and to arrive at substantive decisions. In deliberation, people discuss, ponder, exchange observations and views, reflect upon information and judgments concerning matters of mutual interest, and attempt to persuade each other. Deliberations about risk often include discussions of the role, subjects, methods, and results of *analysis*. Bargaining and mediation are specific deliberative processes, as are debating, consulting, and commenting.

Hazard.   An act or phenomenon that has the potential to produce harm or other undesirable consequences to humans or what they value. Hazards may come from physical phenomena (such as radioactivity, sound waves, magnetic fields, fire, floods, explosions), chemicals (ozone, mercury, dioxins, carbon dioxide, drugs, food additives), organisms (viruses, bacteria), commercial products (toys, tools, automobiles), or human behavior (drunk driving, firing guns). Hazards can also come from information (e.g., information that a person carries a gene that increases susceptibility to cancer may expose the person to job discrimination or increased insurance costs).

Interested parties.   People, groups, or organizations that decide to become informed about and involved in a *risk characterization* or decision-making process. Interested parties may or may not also be *affected parties*.

Problem formulation.   An activity in which public officials, scientists, and *interested and affected parties* clarify the nature of the choices to be considered, the attendant hazards and risks, and the knowledge needed to inform the choices. Problem formulation sets the agenda for the other steps leading to a *risk characterization*: process design, selection of options and outcomes to consider, gathering and interpreting information, and synthesis.

Risk.   A concept used to give meaning to things, forces, or circumstances that pose danger to people or to what they value. Descriptions of risk are typically stated in terms of the likelihood of harm or loss from a *hazard* and usually include: an identification of what is "at risk" and may be harmed or lost (e.g., health of human beings or an ecosystem, personal property, quality of life, ability to carry on an economic activity); the hazard that may occasion this loss; and a judgment about the likelihood that harm will occur.

Risk analysis.   The application of methods of *analysis* to matters of risk. Its aim is to increase understanding of the substantive qualities, seriousness, likelihood, and conditions of a hazard or risk and of the options for managing it.   Although risk analysis is sometimes conceived to be relevant only to gathering, interpreting, and summarizing information about certain possible consequences of a hazard, analysis has other uses in risk characterization.

Risk characterization .   A synthesis and summary of information about a *hazard* that addresses the needs and interests of decision makers and of interested and affected parties.   Risk characterization is a prelude to decision making and depends on an iterative, analytic-deliberative process.

References

Albert, R.
   1994   Carcinogen risk assessment in the U.S. Environmental Protection Agency. *Critical Reviews in Toxicology* 24(1):75-85.
Anderson, E.
   1988   Values, risks, and market norms. *Philosophy and Public Affairs* 17:54-65.
Anderson, R.
   1986   Public participation in hazardous waste facility location decisions. *Journal of Planning Literature* 1(2):145-161.
Armour, A.
   1995   The citizens' jury model of public participation: A critical examination. Pp. 175-187 in O. Renn, T. Webler, and P. Wiedemann, eds., *Fairness and* Competence in Citizen Participation: Evaluating Models for Environmental *Discourse*. Dordrecht, Netherlands: Kluwer Academic Publishers.
Arnstein, S.
   1969   A ladder of citizen participation. *Journal of the American Institute of Planners* 35:216-224.
Arrow, K.J., M.L. Cropper, G.C. Eads, R.W. Hahn, L.B. Lave, R.G. Noll, P.R. Portney, M. Russell, R. Schmalensee, V.K. Smith, and R.N. Stavins
   1996   *Benefit-Cost Analysis in Environmental, Health, and Safety Regulation: A Statement of Principles.* Washington, D.C.: American Enterprise Institute, The Annapolis Center, and Resources for the Future.
A.T. Kearney, Incorporated
   1993   WTI Phase II Risk Assessment Project Plan. EPA ID Number OHD980613541. Prepared for U.S. Environmental Protection Agency Region 5, Chicago, Illinois.
Bannerman, D.G.
   1987   Testimony, National Electrical Manufacturers Association, before the House Committee on Energy and Commerce, Subcommittee on Transportation, Tourism, and Hazardous Materials. December 9.

Barber, B.
   1984   *Strong Democracy: Participatory Politics for a New Age.* Berkeley: University of
           California Press.
Bartell, S.M., R.H. Garner, and R.V. O'Neill
   1992   *Ecological Risk Estimation.* Boca Raton, Fla.: Lewis Publishers.
Baughman, M.
   1995   Mediation. Pp. 253-260 in O. Renn, T. Webler, and P. Wiedemann, eds., *Fairness
           and Competence in Citizen Participation: Evaluating Models for Environmental Dis-
           course.* Dordrecht, Netherlands: Kluwer Academic Publishers.
Beach, L.R.
   1990   *Image Theory: Decision Making in Personal and Organizational Contexts.* New York:
           Wiley.
Beck, M.B.
   1987   Water quality modeling: A review of the analysis of uncertainty. *Water Resources
           Research* 23:1393-1442.
Bella, D.A.
   1987   Engineering and erosion of trust. *Journal of Professional Issues in Engineering*
           113:117-129.
Bingham, G.
   1986   *Resolving Environmental Disputes: A Decade of Experience.* Washington, D.C.: World
           Wildlife Fund.
Blush, S., and T. Heitman
   1995   *Train Wreck Along the River of Money.* Washington, D.C.: U.S. Senate Committee
           on Energy and Natural Resources.
Bohnenblust, H., and T. Schneider
   1984   Risk Appraisal: Can It Be Improved by Formal Models? Paper presented at the
           Annual Meeting of the Society for Risk Analysis, Knoxville, Tenn.
Bostrom, A., B. Fischhoff, and M.G. Morgan
   1992   Characterizing mental models of hazardous processes: A methodology and an
           application to radon. *Journal of Social Issues* 48(4):85-100.
Bottcher, A.B., and F.T. Izuno, eds.
   1994   *Everglades Agricultural Area (EAA). Water, Soil, Crop, and Environmental Manage-
           ment.* Gainesville: University Press of Florida.
Bradbury, J.A.
   1989   The policy implications of differing concepts of risk. *Science, Technology, & Hu-
           man Values* 14:380-399.
Brody, J.G., and R.A. Rosen
   1994   Apples and oranges: Using multi-attribute analysis in a collaborative process to
           address value conflicts in electric facility siting. *NRRI Quarterly Bulletin* (Na-
           tional Regulatory Research Institute) 15(4):629-643.
Brown, J.
   1989   *Environmental Threats: Social Science Approaches to Public Risk Perception.* London,
           U.K.: Belhaven.
Brown, P.
   1990   Popular epidemiology: Community response to toxic-waste-induced disease. Pp.
           77-85 in P. Conrad and R. Kern, eds., *The Sociology of Health and Illness in Critical
           Perspective.* New York: St. Martin's Press.
Browner, C.M.
   1995   Policy for Risk Characterization at the U.S. Environmental Protection Agency.
           Washington, D.C.: U.S. Environmental Protection Agency.

Brunner, R.D.
  1991   Global climate change: Defining the policy problem. *Policy Sciences* 24:291-311.
Bullard, R.
  1990   *Dumping in Dixie: Race, Class and Environmental Quality.* Boulder, Colo.: Westview Press.
Bullard, R., and B. Wright
  1989   Toxic waste and the African-American community. *Urban League Review* 13:67-75.
  1992   *Confronting Environmental Racism: Voices from the Grassroots.* Boston, Mass.: South End Press.
Burke, T.
  1995   Back to the future: Rediscovering the role of public health in environmental decision making. In C.R. Cowthern, ed., *Handbook for Environmental Risk Decision Making: Values, Perceptions, and Ethics.* Boca Raton, Fla.: Lewis Publishers.
Burke, T.A., N.L. Tran, J.S. Roemer, and C.J. Henry, eds.
  1993   *Regulating Risk: The Science and Politics of Risk.* Washington, D.C.: International Life Sciences Institute.
California Department of Food and Agriculture
  1994   The Exotic Fruit Fly Eradication Program Using Aerial Application of Malathion and Bait. Final Programmatic Environmental Impact Report. State Clearinghouse Number 91043018. April.
California Environmental Protection Agency
  1992   Toward the 21st Century: Planning for the Protection of California's Environment. The Comparative Risk Project Workplan. Sacramento: California Environmental Protection Agency.
  1994   Toward the 21st Century: Planning for the Protection of California's Environment. California Comparative Risk Project Final Report. Sacramento: California Environmental Protection Agency.
Carey, J.R.
  1991   Establishment of the Mediterranean fruit fly in California. *Science* 253:1369-1373.
  1994   The Mediterranean fruit fly invasion of Southern California. Pp. 71-91 in J.G. Morse, R.L. Metcalf, J.R. Carey, R.V. Dowell, eds., *Proceedings: The Medfly In California: Defining Critical Research, 9-11 November.* Riverside: University of California Center for Exotic Pest Research.
Chaloner, K.
  1996   The elicitation of prior distributions. Pp. 141-156 in D. Berry and D. Stangl, eds., *Bayesian Biostatistics.* New York: Marcel Dekker.
Checkoway, B.
  1981   The politics of public hearings. *The Journal of Applied Behavioral Science* 17(4):566-581.
Chess, C.
  1989   *Drafting Water Quality Regulations: A Case Study in Public Participation.* New Brunswick, N.J.: Center for Environmental Communication, Rutgers University.
Chess, C., and B.J. Hance
  1994   *Communicating with the Public: Ten Questions Environmental Managers Should Ask.* New Brunswick, N.J.: Center for Environmental Communication, Rutgers University.
Chess, C., and K. Salomone
  1992   Rhetoric and reality: Risk communication in government agencies. *Journal of Environmental Education* 23:28-33.

Chess, C., K. Salomone, and B.J. Hance
    1995   Improving risk communication in government: Research priorities. *Risk Analysis*
            15(1):127-135.
Chess C., K. Salomone, B.J. Hance, and A. Saville
    1995   Results of a national symposium on risk communication: Next steps for govern-
            ment agencies. *Risk Analysis* 15(2):115-125.
Chess, C., M. Tamuz, and M. Greenberg
    1995   Organizational learning about environmental risk communication: The case of
            Rohm and Haas' Bristol plant. *Society and Natural Resources* 8:57-66.
Clapp, R., P. deFur, E. Silbergeld, and P. Washburn
    1995   EPA on the right track. *Environmental Science & Technology* 29(1):29-30.
Clarke, L.
    1988   Politics and bias in risk assessment. *The Social Science Journal* 15:155-165.
    1993   The disqualification heuristic: When do organizations misperceive risk? *Research
            in Social Problems and Public Policy* 5:289-312.
Clarke, L., and J.F. Short, Jr.
    1993   Social organization and risk: Some current controversies. *Annual Review of Sociol-
            ogy*, 19:375-399
Claus, F.
    1995   The Varresbecker Bach participatory process: The model of citizen initiatives. Pp.
            193-206 in O. Renn, T. Webler and P. Wiedemann, eds., *Fair and Competent Citizen
            Participation: Evaluating New Models for Environmental Discourse*. Dordrecht, Neth-
            erlands: Kluwer Academic Publishers.
Clifford, F.
    1994   Cal/EPA's newest hazard: Risks to peace of mind. *Los Angeles Times,* June 11.
Clemen, R.T.
    1990   *Making Hard Decisions: An Introduction to Decision Analysis*. Belmont, Calif.:
            Duxbury Press.
Cohen, J.
    1994   Toxic dispute costs Stanford $1 million. *Science* 266:213.
Cohen, S.M., and L.B. Ellwein
    1995a   A biological theory for carcinogenesis. In S. Olin, W. Farland, C. Park, L.
            Rhomberg, R. Scheuplein, T. Starr, and J. Wilson, eds., *Low-Dose Extrapolation of
            Cancer Risks: Issues and Perspectives*. Washington, D.C.: ILSI Press.
    1995b   Cell proliferation in carcinogenesis. *Science* 249:1007-1011.
Cohrssen, J., and V.T. Covello
    1989   *Risk Analysis: A Guide to Principles and Methods for Analyzing Health and Environ-
            mental Risks*. Springfield, Va.: National Technical Information Service.
Cole, G.A., and S.B. Whithey
    1981   Perspective on risk perception. *Risk Analysis* 1:143-163.
Commission for Racial Justice
    1987   *Toxic Wastes and Race in the United States: A National Report on the Racial and Socio-
            economic Characteristics of Communities with Hazardous Waste Sites*. New York:
            Public Data Access.
Committee on Science, Engineering, and Public Policy
    1992   *Policy Implications of Greenhouse Warming*. Washington, D.C.: National Academy
            Press.
Cormick, G.
    1987   The myth, the reality, and the future environmental mediations. In R. Lake, ed.,
            Resolving Locational Conflict. New Jersey: Center for Urban Policy Research,
            Rutgers University.

Cox, D.C., and P. Baybutt
   1981   Methods for uncertainty analysis: A comparative study. *Risk Analysis* 1:251-258.
Creighton, J.
   1980   *Public Involvement Manual: Involving the Public in Water and Power Resources Decisions.* Washington, D.C.: U.S. Government Printing Office.
Creighton, J., and J. Delli Priscolli
   1983   *Public Involvement Techniques.* Washington, D.C.: U.S. Army Corps of Engineers.
Cropper, M.L., S.K. Aydede, and P.R. Portney
   1994   Preferences for life saving programs: How the public discounts time and age. *Journal of Risk and Uncertainty* 8:243-246.
Cropper, M.L., and W.E. Oates
   1992   Environmental economics: A survey. *Journal of Economic Literature* 30:675-740.
Crosby, N.
   1995   Citizen juries: One solution for difficult environmental problems. Pp. 157-174 in O. Renn, T. Webler, and P. Wiedemann, eds., *Fairness and Competence in Citizen Participation: Evaluating Models for Environmental Discourse.* Dordrecht, Netherlands: Kluwer Academic Publications.
Crosby, N., J. Kelly, and P. Schaeffer
   1986   Citizen panels: A new approach to citizen participation. *Public Administration Review* 46:170..
Crouch, E., and R. Wilson
   1982   *Risk/Benefit Analysis.* Cambridge, Mass.: Ballinger Publishing Company.
Crowfoot, J.E., and J.M. Wollendeck
   1990   *Environmental Disputes: Community Involvement in Conflict Resolution.* Washington, D.C.: Island Press.
Dake, K.
   1991   Orienting dispositions in the perception of risk: An analysis of contemporary world views and cultural biases. *Journal of Cross-Cultural Psychology* 22:61-82.
Dakins, M.E., J.E. Toll, and M.J. Small
   1994   Risk-based environmental remediation: Decision framework and role of uncertainty. *Environmental Toxicology & Chemistry* 13(12):1907-1915.
Davis, S.M., and J.C. Ogden
   1994   *Everglades: The Ecosystem and Its Restoration.* Delray Beach, Fla.: St. Lucie Press.
Dean, W., and P. Thompson.
   1995   *The Varieties of Risk* (ERC 95-3). Eco-Research Chair, Environmental Risk Management. Calgary: University of Alberta.
Dickey, J.M.
   1980   Beliefs about beliefs: A theory of stochastic assessments of subjective probabilities. In D.J.L.M. Bernardo, M.H. DeGroot, and A. Smith, eds. *Case Studies in Bayesian Biostatistics.* Valencia, Spain: University Press.
Dienel, P.
   1989   Contributing to social decision methodology: Citizen reports on technological projects. Pp. 133-152 in C. Vlek and G. Cvetkovich, eds., *Social Decision Methodology for Technological Projects.* Dordrecht, Netherlands: Kluwer Academic Press.
Dienel, P., and O. Renn
   1995   A gate to fractal mediation. Pp 118-124 in O. Renn, T. Webler, and P. Wiedemann, eds., *Fairness and Competence in Citizen Participation: Evaluating Models for Environmental Discourse.* Dordrecht, Netherlands: Kluwer Academic Publishers.
Dietz, T.
   1987   Theory and method in social impact assessment. *Sociological Inquiry* 57(Winter): 54-69.

1994   What should we do? Human ecology and collective decision making. *Human Ecology Review* 1(Summer/Autumn):301-309.

Dietz, T., and R. Rycroft
1987   *The Risk Professionals.* New York: Russell Sage Foundation.

Dietz, T., P.C. Stern, and R. Rycroft
1989   Definitions of conflict and the legitimation of resources: The case of environmental risk. *Sociological Forum* 4:47-70.

Doniger, D.
1986   Negotiated rulemaking at EPA: The examples of wood stove emissions and truck engine emissions. In Standing Committee on Environmental Law, ed., *The Private Assumption of Previously Public Responsibilities: The Expanding Role of Private Institutions in Public Environmental Decisionmaking.* Washington, D.C.: American Bar Association.

Douglas, M.S.
1947   *The Everglades: River of Grass.* Miami, Fla.: Banyan Books.

Douglas, M., and A. Wildavsky
1982   *Risk and Culture.* Berkeley: University of California Press.

Dryzek, J., and S. Hunter
1987   Environmental mediation for international problems. *International Studies Quarterly* 31:87-102.

Dwyer, J.
1990   The pathology of symbolic legislation. *Ecology Law Quarterly* 17:233-316.

Edwards, W., and J.R. Newman
1982   *Multiattribute Evaluation.* Paper No. 26. New York: Sage.

Edwards, W., and D. von Winterfeldt
1987   Public values in risk debates. *Risk Analysis* 7:141-158.

Ehrmann, J., and B. Stinson
1994   Human health impact assessment: The link with alternative dispute resolution. *Environmental Impact Assessment Review* 14:517-526.

Ell, K.O., and R.H. Nishimoto
1989   Coping resources in adaptation to cancer: Socioeconomic and racial differences. *Social Service Review* 63:443-446.

Elliot, M.
1984   Improving community acceptance of hazardous waste facilities through alternative systems for mitigating and managing risk. *Hazardous Waste* 1:397-410.

Ellis, R.D.
1993   Quantifying distributive justice: An approach to environmental and risk-related public policy. *Policy Sciences* 26:99-103.

English, M.R.
1992   *Siting Low-Level Radioactive Waste Disposal Facilities: The Public Policy Dilemma.* New York: Quorum.

English, M.R., A.K. Gibson, D.L. Feldman, and B.E. Tonn
1993   *Stakeholder Involvement: Open Processes for Reaching Decisions about the Future Uses of Contaminated Sites.* Final Report. Knoxville: Waste Management Research and Education Institute, University of Tennessee.

Environ Dioxin Risk Characterization Expert Panel
1995   EPA assessment not justified. *Environmental Science & Technology* 29(1):31-32.

Evans, N., and C. Hope
1984   *Nuclear Power: Futures, Costs, and Benefits.* Cambridge, England: Cambridge University Press.

Ferguson, S.
1986 Industry-environmentalist negotiation: The FIFRA experience. Pp. 11-13 in Standing Committee on Environmental Law, ed., *The Private Assumption of Previously Public Responsibilities: The Expanding Role of Private Institutions in Public Environmental Decisionmaking.* Washington, D.C.: American Bar Association.

Fessenden-Raden, J., J.M. Fitchen, and J.S. Heath
1987 Providing risk information in communities: Factors influencing what is heard and accepted. *Science Technology and Human Values* 12:94-101.

Finkel, A.M.
1990 *Confronting Uncertainty in Risk Management—A Guide for Decision-Makers.* Washington, D.C.: Center for Risk Management, Resources for the Future.

1991 Edifying presentation of risk estimates: Not as easy as it seems. *Journal of Policy Analysis and Management* 10:296-303.

1994 The case for "plausible conservatism" in choosing and altering defaults. Pp. 601-627 in National Research Council, *Science and Judgment in Risk Assessment.* Washington, D.C.: National Academy Press.

Finkel, A.M., and J.S. Evans
1987 Evaluating the benefits of uncertainty reduction in environmental health risk management. *Journal of the Air Pollution Control Association* 37:1164-1171.

Finkel, A.M., and D. Golding, eds.
1995 *Worst Things First: The Debate Over Risk-Based National Environmental Priorites.* Washington, D.C.: Resources for the Future.

Finsterbusch, K., and C.P. Wolf, eds.
1981 *Methodology of Social Impact Assessment,* second edition. Stroudsburg, Pa.: Hutchinson and Ross.

Finsterbusch, K., L. Llewellyn, and C.P. Wolf, eds.
1984 *Social Impact Assessment Methods.* Beverly Hills, Calif.: Sage.

Fiorino, D.J.
1988 Regulatory negotiations as a policy process. *Public Administration Review* 48:764-772.

1989 Environmental risk and democratic process: A critical review. *Columbia Journal of Environmental Law* 14:501-547.

1990 Citizen participation and environmental risk: A survey of institutional mechanisms. *Science, Technology, and Human Values* 15:226-243.

1991 Dimensions of negotiated rulemaking: practical constraints and theoretical implications. Pp. 127-139 in S. Nagel and M. Mills, eds., *Systematic Analysis in Dispute Resolution.* New York: Quorum Books.

1995 Regulatory negotiation as a form of public participation. Pp. 223-237 in O. Renn, T. Webler, and P. Wiedemann, eds., *Fairness and Competence in Citizen Participation: Evaluating Models for Environmental Discourse.* Dordrecht, Netherlands: Kluwer Academic Publishers.

Fischer, F.
1990 *Technocracy and the Politics of Expertise.* Newbury Park, Calif.: Sage Publications.

1993 Citizen participation and the democratization of policy expertise: From theoretical inquiry to practical cases. *Policy Sciences* 26:165-187.

Fischhoff, B.
1984 Setting standards: A systematic approach to managing public health and safety risks. *Management Science* 30:823-843.

1989 Risk: A guide to controversy. Pp. 211-319 in National Research Council, *Improving Risk Communication.* Washington, D.C.: National Academy Press.

1991   The Psychology of Risk Characterization. Paper presented at the Workshop on Risk Characterization sponsored by American Industrial Health Association, Environmental Protection Agency, and Resources for the Future. September 26-27, Washington, D.C.

1994   Acceptable risk: A conceptual proposal. *Risk: Health, Safety and the Environment* 5:1-28.

1995   Ranking risks. *Risk: Health, Safety and the Environment* 6:191-202.

Fischhoff, B., A. Bostrom, and M.J. Quadrel
1993   Risk perception and communication. *Annual Review of Public Health* 14:183-203.

Fischhoff, B., S. Lichtenstein, P. Slovic, S. Derby, and R. Keeney
1981   *Acceptable Risk*. New York: Cambridge University Press.

Fischhoff, B., S.R. Watson, and C. Hope
1984   Defining risk. *Policy Sciences* 17:123-139.

Fisher, A., M. Pavlova, and V. Covello, eds.
1991   *Evaluation and Effective Risk Communication Workshop Proceedings*. Interagency Task Force on Environmental Cancer and Heart and Lung Disease, EPA/600/9-90/-54 (January). Washington, D.C.: U.S. Environmental Protection Agency.

Flynn, J., and P. Slovic
1993   Nuclear wastes and public trust. *Forum for Applied Research and Public Policy* 8:92-100.

Flynn, J., P. Slovic, and C.K. Mertz
1994   Gender, race, and perception of environmental health risks. *Risk Analysis* 14(6):1101-1108.

Foran, J., B. Goldstein, J. Moore, and P. Slovic
1995   Predicting Future Sources of Mass Toxic Tort Litigation. Unpublished manuscript. Risk Science Institute, Washington, D.C.

Fort, R., R. Rosenman, and W. Budd
1993   Perception costs and NIMBY. *Journal of Environmental Management* 38:185-200.

Freeze, R.A., J.W. Massmann, L. Smith, T. Sperling, and B. James
1990   Hydrogeological decision analysis: 1. A framework. *Ground Water* 28(5):738-766.

Freudenburg, W.
1988   Perceived risk, real risk: Social science and the art of probabilistic risk assessment. *Science* 241:44-49.

1992   Nothing recedes like success? Risk analysis and the organizational amplification of risks. *Risk: Issues in Health and Safety* 3:1-38.

Funk, W.
1987   When smoke gets in your eyes: Regulatory negotiation and the public interest. *Environmental Law* 18:55-98.

Funtowicz, S.O., and J.R. Ravetz
1992   Three types of risk assessment and the emergence of post-normal science. Pp. 251-274 in S. Krimsky and D. Golding, eds., *Social Theories of Risk*. Westport, Conn.: Praeger.

Genest, C., and M.J. Schervish
1985   Modeling expert judgements for Bayesian updating. *The Annals of Statistics* 13(3):1198-1212.

Gough, M., and J.D. Wilson
1994   Understanding the relationship between science and risk analysis. *AIHC Journal* 2(1):12-14.

Graham, J.
1985   The failure of agency-forcing: The regulations of airborne carcinogens under Section 112 of the Clean Air Act. *Duke Law Journal* 1985:100-150.

Graham, J.D., L.C. Green, and M.J. Roberts
  1988   *In Search of Safety: Chemicals and Cancer Risk.* Cambridge, Mass.: Harvard University Press.

Gray, B.
  1989   *Collaborating: Finding Common Ground for Multiparty Problems.* San Francisco, Calif.: Jossey-Bass.

Greenberg, M.
  1993   Proving environmental inequity in siting locally unwanted land uses. *Risk: Issues in Health and Safety* 4:235-252.
  1995   Separate and not equal: Health-environmental risk and economic-social impacts in remediating hazardous waste sites. In S.K. Majumdar, F.J. Brenner, L.M. Rosenfeld, and E.W. Miller, eds., *Environmental Contaminants and Health.* Easton: Pennsylvania Academy of Sciences (Lafayette College).

Greenberg, M., and J. Hughes
  1992   The impact of hazardous waste superfund sites on the value of homes sold in New Jersey. *Annals of Regional Science* 26:147-153.
  1993   Impact of hazardous waste sites on property values and land use: Tax assessors' appraisal. *The Appraisal Journal* 42-51.

Gregory, R., J. Flynn, and P. Slovic
  1995   Technological stigma. *American Scientist* 83:220-223.

Gregory, R., S. Lichtenstein, and P. Slovic
  1993   Valuing Environmental Resources: A Constructive Approach. *Journal of Risk and Uncertainty* 7:177-197.

Gregory, R., S. Lichtenstein, and D.G. MacGregor
  1993   The role of past states in determining reference points for policy decisions. *Organizational Behavior and Human Decision Processes* 55:195-206.

Griesmeyer, J.M., and D. Okrent
  1981   Risk management and decision rules for light water reactors. *Risk Analysis* 1:121-136.

Grumbine, R.E.
  1994   What is ecosystem management? *Conservation Biology* (8)1:27-38.

Grumbly, T.
  1996   Letter to "Site Specific Advisory Board Participants" January 29, 1996. (DOE F 1291.3) U.S. Department of Energy, Washington, D.C.

Habicht, H. F.
  1992   Guidance on risk characterization for risk managers and risk assessors. Published as pp. 351-374 in National Research Council, *Science and Judgment in Risk Assessment.* Washington, D.C.: National Academy Press, 1994.

Hadden, S.
  1989   *A Citizen's Right-to-Know: Risk Communication and Public Policy.* Boulder, Colo.: Westview Press.
  1995   Regulatory negotiation as citizen participation: A critique. Pp. 239-252 in O. Renn, T. Webler, and P. Wiedemann, eds., *Fair and Competent Citizen Participation: Evaluating New Models for Environmental Discourse.* Dordrecht, Netherlands: Kluwer Academic Publishers.

Hanford Future Sites Working Group.
  1992   The Future for Hanford: Uses and Cleanup. U.S. Department of Energy, Richland, Wash.

Harris, J.
  1990   Environmental policy making: Act now or wait for more information? Pp. 107-133 in P.B. Hammond and R. Coppock, eds., *Valuing Health Risks, Costs, and Ben-*

efits for Environmental Decision Making. Washington, D.C.: National Academy Press.

Harter, P.
1982   *Negotiating regulations: A cure for the malaise?* Georgetown Law Review 71:1-118.

Harwell, M.A., W. Cooper, and R. Flaak
1992   Prioritizing ecological and human welfare risks from environmental stresses. *Environmental Management* 16:451-464.

Harwell, M., J. Gentile, B. Norton, and W. Cooper
1994   Issue paper on ecological significance. Chapter 2 in *Ecological Risk Assessment Issue Papers.* Washington, D.C.: U.S. Environmental Protection Agency.

Harwell, M.A., C.C. Harwell, D.A. Weinstein, and J.R. Kelly
1990   Characterizing ecosystem responses to stress. Pp. 91-115 in E.B. Cowling, A.I. Breymeyer, A.S. Phillips, S.I. Auerback, A.M. Bartuska, and M.A. Harwell, eds., *Ecological Risks: Perspectives from Poland and the United States.* Washington, D.C.: National Academy Press.

Harwell, M.A. and T.C. Hutchinson, with W.P. Cropper, Jr., C.C. Harwell, and H.D. Grover
1989   *Environmental Consequences of Nuclear War: Volume II Ecological and Agricultural Effects.* SCOPE 28. Second Edition. Chichester, U.K.: John Wiley & Sons.

Harwell, M.A., J.F. Long, A. Bartuska, J.H. Gentile, C.C. Harwell, V. Myers, and J.C. Ogden
in   Ecosystem management to achieve ecological sustainability: The case of South
press   Florida. *Environmental Management.*

Harwell, M.A., and J.F. Long, eds.
1995   *US Man and the Biosphere Human-Dominated Systems Directorate Workshops on Ecological and Societal Issues for Sustainability.* Washington, D.C.: US Man and the Biosphere Program.

Heberlein, T.A.
1976   Some observations on alternative mechanisms for public involvment: The hearing, public opinion poll, the workshop, and the quasi-experiment. *Natural Resources Journal* 16(1):197-213.

Heiman, M.
1990   From "not in my backyard!" to "not in anybody's backyard!" *Journal of the American Planning Association* 56:359-362.

Heimer, C.A.
1988   Social structures, psychology, and the estimation of risk. *Annual Review of Sociology* 14:491-519.

Heising, C.D., and V.P. George
1986   Nuclear financial risk: Economy-wide costs of reactor accidents. *Energy Policy* 14:45-52.

Hester, G., M.G. Morgan, I. Nair, and K. Florig
1990   Small group studies of regulatory decision-making for power-frequency electric and magnetic fields. *Risk Analysis* 10:213-227.

Hilgartner, S.
1985   The political language of risk: Defining occupational health. Pp. 25-66 in D. Nelkin, ed., *The Language of Risk.* Beverly Hills, Calif.: Sage.

Hornstein, D.
1992   Reclaiming environmental law: A normative critique of comparative risk analysis. *Columbia Law Review* 92:562-633.

Howard, R.A.
1966   Decision analysis: Applied statistical decision theory. Pp. 55-71 in D.B. Hertz and J. Melese, eds., *Proceedings of the Fourth International Conference on Operations Research.* New York: Wiley-Interscience.

1968   The foundations of decision analysis. *IEEE Transactions on Systems Science and Cybernetics* SSC-4(3) (September):1-9.

Howard, R.A., and J.E. Matheson, eds.
1984   *Readings on the Principles and Applications of Decision Aanalysis.* Menlo Park, Calif.: Strategic Decisions Group.

Howard, R.A., J.E. Matheson, and D.W. North
1972   The decision to seed hurricanes. *Science* 176:1191-1202.

Houghton, D.
1988   Citizen advisory boards: Autonomy and effectiveness. *American Review of Public Administration* 18:283-296.

Iman, R.L., and J.C. Helton
1988   An investigation of uncertainty and sensitivity analysis techniques for computer models. *Risk Analysis* 8(1):71-90.

Interagency Regulatory Liaison Group, Work Group on Risk Assessment
1979   Scientific bases for identification of potential carcinogens and estimation of risks. *Journal of the NCI* 63(1, July):241-268.

James, B.R., and R.A. Freeze
1993   The worth of data in predicting aquitard continuity in hydrogeological design. *Water Resources Research* 29(7):2049-2065.

James, B.R., and S.M. Gorelick
1994   When enough is enough: The worth of monitoring data in aquifer remediation design. *Water Resources Research* 30(12):3499-3513.

Jasanoff, S.
1986   *Risk Management and Political Culture.* New York: Russell Sage Foundation.
1987a   Contested boundaries in policy-relevant science. *Social Studies of Science* 17(2):195-230.
1987b   Cultural aspects of risk assessment in Britain and the United States. Pp. 359-397 in B.B. Johnson and V.T. Covello, eds., *The Social and Cultural Construction of Risk.* Dordrecht, Netherlands: Reidel.
1987c   EPA's regulation of daminozide: Unscrambling the messages of risk. *Science, Technology, and Human Values* 12:116-124.
1990   American exceptionalism and the political acknowledgment of risk. *Daedalus* 119(4):61-79.
1991   Acceptable evidence in a pluralistic society. Pp. 29-47 in D.G. Mayo and R.G. Hollander, eds., *Acceptable Evidence: Science and Values in Risk Management.* New York: Oxford University Press.
1993   Bridging the two cultures of risk analysis. *Risk Analysis* 13:123-129.

Jefferson Center
1988   *Final Report, Policy Jury and Human Services Committee of the Minnesota Senate.* Minneapolis, Minn.: Jefferson Center.

Jefferys, W.H., and J.O. Berger
1992   Ockham's razor and Bayesian analysis. *American Scientist* 80:64-72.

Johnson, B.B., and V.T. Covello
1987   *The Social and Cultural Construction of Risk.* Dordrecht, Netherlands: Reidel.

Johnson, B.B., and P. Slovic
1995   Presenting uncertainty in health risk assessment: Initial studies of its effects on risk perception and trust. *Risk Analysis* (15)4:485-494.

Johnson, J.
1996   Risk assessment draft gives WTI incinerator clean slate. *Environmental Science and Technology* 30(1):14A-15A.

Kadane, J.B., J.M. Dickey, R.L. Winkler, W.S. Smith, and S.C. Peters
   1980   Interactive elicitation of opinion for a normal linear model. *Journal of American Statistical Association* 75:845-854.
Kahneman, D., and A. Tversky
   1972   Subjective probability: A judgment of representativeness. *Cognitive Psychology* 3:430-454.
   1973   On the psychology of prediction. *Psychological Review* 80(4):237-251.
   1979   Prospect theory: An analysis of decision under risk. *Econometrica* 47:263-289.
Kahneman, D., P. Slovic, and A. Tversky
   1982   *Judgment under Uncertainty: Heuristics and Biases.* New York: Cambridge University Press.
Kaplan, S., and B.J. Garrick
   1981   On the quantitative definition of risk. *Risk Analysis* 1(1):11-28.
Kasperson, R.E.
   1986   Six propositions for public participation and their relevance for risk communication. *Risk Analysis* 6(3):275-281.
Kasperson, R., D. Golding, and S. Tuler
   1992   Social distrust as a factor in siting hazardous facilities and communicating risks. *Journal of Social Issues* 48(4):161-187.
Kasperson, R., O. Renn, P. Slovic, H. Brown, J. Emel, R. Goble, J. Kasperson, and S. Ratick
   1988   The social amplification of risk: A conceptual framework. *Risk Analysis* 8:177-187.
Kathlene, L., and J. Martin
   1991   Enhancing citizen participation: Panel designs, perspectives and policy formation. *Journal of Policy Analysis and Management* 10:46-63.
Keeler, E.B., and S. Cretin
   1983   Discounting of life-saving and other nonmonetary effects. *Management Science* 29:300-306.
Keeney, R.L.
   1980   *Siting Energy Facilities.* New York: Academic Press.
Keeney, R.L., and H. Raiffa
   1976   *Decisions with Multiple Objectives: Preferences and Value Tradeoffs.* New York: John Wiley & Sons.
Keeney, R., D. von Winterfeldt, and T. Eppel
   1990   Eliciting public values for complex policy decisions. *Management Science* 36:1011-1030.
Kelman, S.
   1992   Adversary and cooperationist institutions for conflict resolution in public policy making. *Journal of Policy Analysis and Management* 11:178-206.
Kempton, W.
   1991   Lay perspectives on global climate change. *Global Environmental Change: Human and Policy Dimensions* 1:183-208.
Konheim, C.S.
   1988   Risk communication in the real world. *Risk Analysis* 8:367-373.
Kopp, R., and V.K. Smith
   1993   *Valuing Natural Assets: The Economics of Natural Resource Damage Assessment.* Washington, D.C.: Resources for the Future.
Kraft, M.
   1988   Evaluating technology through public participation: The nuclear waste disposal controversy. Pp. 253-277 in M.E. Kraft und N.J. Vig, eds., *Technology and Politics.* Durham: Duke University Press.

Kraus, N., T. Malmfors, and P. Slovic
    1991   Intuitive toxicology: Expert and lay judgments of chemical risks. *Risk Analysis* 12:215-232.
Krauskopf, K.B.
    1990   Disposal of high-level nuclear waste: Is it possible? *Science* 249:1231-1232.
Kressel, K., and D.G. Pruitt, eds.
    1985   The mediation of social conflict. *Journal of Social Issues* 41(Whole No. 2).
Krimsky, S., and D. Golding, eds.
    1992   *Social Theories of Risk*. Westport, Conn: Praeger.
Krimsky, S., and A. Plough
    1988   *Environmental Hazards: Communicating Risks as a Social Process*. Dover, Mass.: Auburn House.
Krofchick, M.A., J.H. Garrett, Jr., and S.J. Fenves
    1995   A broker for delivering and accessing environmental regulations. Pp. 36-43 in *ASCE Conference on Computing in Civil Engineering* (Atlanta, Ga.) New York: American Society of Civil Engineers.
Kuhn, T.S.
    1970   *The Structure of Scientific Revolutions*. Chicago: University of Chicago Press.
Kunreuther, H.
    1987   Gridlock in environmental insurance: The failure of EIL coverage. *Environment* 29:18-35.
Kunreuther, H., K. Fitzgerald, and T. Aarts
    1993   Siting noxious facilities: A test of the facility credo. *Risk Analysis* 13:301-315.
Kunreuther, H., and J. Linnerooth
    1982   *Risk Analysis and Decision Processes: The Siting of Liquefied Energy Gas Facilities in Four Countries*. Berlin, Germany: Springer Verlag.
Lackey, R.T.
    1994   Ecological risk assessment. *Fisheries* 19(9):14.
    1995   The future of ecological risk assessment. *Human and Ecological Risk Assessment* 1:339-343.
Laird, F.N.
    1989   The decline of deference: The political context of risk communication. *Risk Analysis* 9:543-550.
    1993   Participatory analysis, democracy, and technological decision making. *Science, Technology, and Human Values* 18(3):341-361.
Lave, L.
    1981   *The Strategy of Social Regulation: Decision Frameworks for Policy*. Washington, D.C.: Brookings Institution.
Lawless, E.W.
    1977   *Technology and Social Shock*. New Brunswick, N.J.: Rutgers University Press.
Leigh, J. P.
    1989   Compensating wages for job-related death: The opposing arguments. *Journal of Economic Issues* 23:823-839.
Leroy, D.H., and T.S. Nadler
    1993   Negotiate way out of siting dilemmas. *Forum for Applied Research and Public Policy* 8:102-107.
Levine, A.G.
    1982   *Love Canal: Science, Politics, and People*. Lexington, Mass.: Lexington Books.
Lewis, H.W., R.J. Budnitz, H.J.C. Kouts, F. von Hippel, W. Lowenstein, and F. Zachariasen
    1975   *Risk Assessment Group Report to the U.S. Nuclear Regulatory Commission*. Washington, D.C.: U.S. Nuclear Regulatory Commission.

Lichtenstein, S., and B. Fischhoff
 1977   Do those who know more also know more about how much they know? *Organizational Behavior and Human Performance* 20:159-183.
Lichtenstein, S., P. Slovic, B. Fischhoff, M. Layman, and B. Combs
 1978   Judged frequency of lethal events. *Journal of Experimental Psychology: Human Learning and Memory* 4:551-578.
Light, S.S., and J.W. Dineen
 1994   Water control in the Everglades: A historical perspective. Pp. 47-84 in S.M. Davis and J.C. Ogden, eds., *Everglades: The Ecosystem and Its Restoration*. Delray Beach, Fla.: St. Lucie Press.
Lilienfield, D.
 1991   The silence: The asbestos industry and early occupational cancer research—A case study. *American Journal of Public Health* 81(6):791-800.
Lindell, B., and T. Malmfors
 1994   Comprehending radiation risks. Pp. 17-18 in B. Lindell et al., eds., *Comprehending Radiation Risks: A Report to the IPEA*. Stockholm, Sweden: Riskkollegiet.
Litai, D., D.D. Lanning, and N.C. Rasmussen
 1983   The public perception of risk. Pp. 213-224 in V.T. Covello, W.G. Flamm, J.V. Rodricks, and R.G. Tardiff, eds., *The Analysis of Actual Versus Perceived Risks*. New York: Plenum.
Lynn, F.M.
 1986   The interplay of science and values in assessing and regulating environmental risks. *Science, Technology, and Human Values* 11(2):40-50.
 1987a   Citizen involvement in hazardous waste sites: Two North Carolina success stories. *Environmental Impact Assessment Review* 7:347-36.
 1987b   OSHA's carcinogens standard: Round one on risk assessment models and assumptions. Pp. 345-358 in B.B. Johnson and V.T. Covello, eds., *The Social and Cultural Construction of Risk*. Dordrecht, Netherlands: Reidel.
Lynn, F.M., and G.J. Busenberg
 1995   Citizen advisory committees and environmental policy: What we know, what's left to discover. *Risk Analysis* 15(2):147-162.
Lynn, F.M., and J.D. Kartez
 1995   The redemption of citizen advisory committees: A perspective from critical theory. Pp. 87-101 in O.Renn, T. Webler, and P. Wiedemann, eds., *Evaluating Models for Environmental Discourse*. Dordrecht, Netherlands: Kluwer Academic Publishers.
MacKenzie, D.
 1990   *Inventing Accuracy: A Historical Sociology of Nuclear Missile Guidance.* Cambridge, Mass.: MIT Press.
MacLean, D.
 1995   A Critical Look at Informed Consent. Unpublished manuscript, Department of Philosophy, University of Maryland at Baltimore County.
Mansbridge, J.
 1983   *Beyond Adversarial Democracy.* New York, N.Y.: John Wiley.
 1990   *Beyond Self Interest.* Chicago, Ill.: University of Chicago Press.
Martin, B.
 1989   The sociology of the fluoridation controversy: A reexamination. *Sociological Quarterly* 30:59-76.
Mazmanian, D.A, and J. Nienaber
 1979   *Can Organizations Change?* Washington, D.C.: Brookings Institution.

Mazur, A.
  1981   *The Dynamics of Technical Controversy.* Washington, D.C.: Communications Press.
  1984   The journalist and technology: Reporting about Love Canal and Three Mile Island. *Minerva* 22:45-66.
McClellan, R.O. and D. North
  1994   Making full use of scientific information in risk assessment. Pp. 629-640 in National Research Council, *Science and Judgement in Risk Assessment.* Washington, D.C.: National Academy Press.
McClelland, G.H., W.D. Schulze, and B. Hurd
  1990   The effect of risk beliefs on property values: A case study of a hazardous waste site. *Risk Analysis* 10(4):485-497.
McNeil, B.J., S.G. Pauker, H.C. Sox, and A. Tversky
  1982   On the elicitation of preferences for alternative therapies. *New England Journal of Medicine* 306:1259-1262.
Michael, M.
  1992   Lay discourses of science: Science-in-general, science-in-particular and self. *Science, Technology, and Human Values* 17(3):313-333.
Milbrath, L.
  1981   Citizen surveys as citizen participation. *Journal of Applied Behavioral Science* (17)4:478-496.
Morgan, M.G.
  1981   Choosing and managing technology-induced risk. *IEEE Spectrum* 18 (December):53-60.
Morgan, M.G., and M. Henrion
  1990   *Uncertainty: A Guide to Dealing with Uncertainty in Quantitative Risk and Policy Analysis.* Cambridge, U.K.: Cambridge University Press.
Morgan, M.G., B. Fischhoff, A. Bostrom, L. Lave, and C. Atman
  1992   Communicating risk to the public: First learn what people know and believe. *Environmental Science and Technology* 26:2049-2056.
Morgan, M.G., S.C. Morris, M. Henrion, D.A.L. Amaral, and W.R. Rish
  1984   Technical uncertainty in quantitative policy analysis—A sulfur air pollution example. *Risk Analysis* 4(3):201-216.
Morse, P.M., and G.E. Kimball
  1951   *Methods of Operations Research.* New York: Technology Press of Massachusetts Institute of Technology and John Wiley and Sons.
Nakamura, R.T., T. W. Church, Jr., and P.J. Cooper
  1991   Environmental dispute resolutions and hazardous waste clean-up: A cautionary take of policy implementation. *Journal of Policy Analysis and Management* 10:204-221.
National Research Council
  1983   *Risk Assessment in the Federal Government: Managing the Process.* Washington, D.C.: National Academy Press.
  1989   *Improving Risk Communication.* Committee on Risk Perception and Communication. Washington, D.C.: National Academy Press.
  1993a   *Issues in Risk Assessment.* Committee on Risk Assessment Methodology. Washington, D.C.: National Academy Press.
  1993b   *Workload Transition: Implications for Individual and Team Performance.* B. M. Huey and C.D. Wickens, eds., Committee on Human Factors. Washington, D.C.: National Academy Press.

1994a   *Science and Judgment in Risk Assessment.* Committee on Risk Assessment of Hazardous Air Pollutants., Board on Environmental Studies and Toxicology. Washington, D.C.: National Academy Press.

1994b   *Building Consensus Through Risk Assessment and Management of the Department of Energy's Environmental Remediation Program.* Washington, D.C.: National Academy Press.

1994c   *Ranking Hazardous-Waste Sites for Remedial Action.* Committee on Remedial Action Priorities for Hazardous Waste Site, Board on Environmental Studies and Toxicology. Washington, D.C.: National Academy Press.

1995   *Technical Bases for Yucca Mountain Standards.* Committee on Technical Bases for Yucca Mountain Standards, Board on Radioactive Waste Management. Washington, D.C.: National Academy Press.

Nelkin, D.
1989   Communicating technological risk: The social construction of risk perception. *Annual Review of Public Health* 10:95-113.

North, D.W.
1995   Use of expert judgment on cancer dose-response: Probabilistic assessment and plans for application to dieldrin. Pp. 275-287 in S. Olin et al., eds., *Low-Dose Extrapolation of Cancer Risks: Issues and Perspectives.* Washington, D.C.: International Life Sciences Institute.

North, D.W., F. Selker, and T. Guardino
1994   The value of research on health effects of ingested inorganic arsenic. Pp. 1-20 in W. Chappell, ed., *Proceedings of the International Conference on Arsenic Exposure and Health Effects, Society for Environmental Geochemistry and Health.* Northwood, U.K.: Science Reviews Ltd.

Nuclear Waste Technical Review Board
1993   *Special Report to Congress and the Secretary of Energy* (March). Arlington, Va.: Nuclear Waste Technical Review Board.

1995   *Report to Congress and the Secretary of Energy* (March). Arlington, Va.: Nuclear Waste Technical Review Board.

O'Brien, M.
1995   A proposal to address, rather than rank, environmental problems. Pp. 87-105 in A.M. Finkel and D. Golding, eds., *Worst Things First: The Debate over Risk-Based National Environmental Priorities.* Washington, D.C.: Resources for the Future.

Office of Cancer Communications, National Cancer Institute
1989   *Making Health Communication Programs Work: A Planners Guide.* (89-1493) Washington: U.S. Department of Health and Human Services.

Olsen, M.E., B.D. Melber, and D.J. Merwin
1981   Methodology for conducting social impact assessments using quality of social life indicators. Pp. 48-78 in K. Finsterbusch and C.P. Wolf, eds., *Methodology of Social Impact Assessment.* Second edition. Stroudsburg, Pa.: Hutchinson and Ross.

Otway, H.
1992   Public wisdom, expert fallibility: Toward a contextual theory of risk. Pp. 215-228 in S. Krimsky and D. Golding, eds., *Social Theories of Risk:* Westport, Conn.: Praeger.

Oreskes, N., K. Shrader-Frechette, and K. Belitz
1994   Verification, validation, and confirmation of numerical models in the earth sciences. *Science* 263:641-646.

Ozawa, C.P.
1991   *Recasting Science: Consensual Procedures in Public Policy Making.* Boulder, Colo.: Westview Press.

Paté-Cornell, E. and P.S. Fischbeck
   1993   PRA as a management tool: Organizational factors and risk-based priorities for
           the maintenance of the tiles of the space shuttle orbiter. *Reliability Engineering and
           Systems Safety* 40:239-257.
Paustenbach, D.
   1989   *The Risk Assessment of Environmental and Human Health Hazards: A Textbook of Case
           Studies.* New York: Wiley.
Peelle, E.
   1979   Mitigating community impacts of energy development. *Nuclear Technology*
           44(June):132-140.
Peelle, E., S.A. Carnes, E.D. Copenhaver, J.H. Sorensen, E.J. Soderstrom, J.H. Reed, and D.J.
Bjornstad
   1983   Incentives and nuclear waste siting: Prospects and Constraints. *Energy Systems
           and Policy* 7(4):329.
Peelle, E., and R. Ellis
   1987   Hazardous waste management outlook. *Forum for Applied Research and Public
           Policy* 2(3):68-77.
Perritt, H.
   1986   Negotiated rulemaking in practice. *Journal of Policy Analysis and Management*
           5:482-495.
Perrow, C.
   1984   *Normal Accidents: Living with High Risk Technologies.* New York: Basic Books.
Peters, E., and P. Slovic
   1995   *The Role of Affect and Worldviews as Orienting Dispositions in the Perception and
           Acceptance of Nuclear Power.* Eugene, Oreg.: Decision Research.
Peterson, C., and A.J. Stunkard
   1989   Personal control and health promotion. *Social Science and Medicine* 28:819-828.
Pflugh, K.K.
   no   The Role of Risk Communication Planning in the Release of the Oral Rabies
   date   Vaccine in New Jersey: An Evaluation. New Jersey Department of Environmen-
           tal Protection and Energy, Division of Science and Research.
Pijawka, K.D., and A.H. Mushkatel
   1992   Public opposition to the siting of the high-level nuclear waste repository: The
           importance of trust. *Policy Studies Review* 10:180-194.
Pidgeon, N., C. Hood, D. Jones, B. Turner, and R. Gibson
   1992   Risk perception. Pp. 89-134 in Royal Society Study Group, ed., *Risk: Analysis,
           Perception and Management.* London, U.K.: The Royal Society.
Pierce, J., and H. Doerksen
   1976   Citizen advisory committees: The impact of recruitment on representation and
           responsiveness. In J. Pierce and H. Doerksen, eds., *Water Politics and Public In-
           volvement.* Ann Arbor, Mich.: Science Publishers.
Plough, A., and S. Krimsky
   1987   The emergence of risk communication studies: Social and political context. *Sci-
           ence, Technology, & Human Values* 12:4-10.
Presidential Commission on the Space Shuttle *Challenger* Accident
   1986   *Report of the Presidential Commission on the Space Shuttle* Challenger *Accident.* Wash-
           ington, D.C.: U.S. National Aeronautics and Space Administration.
President's Commission on the Accident at Three Mile Island
   1979   *Reports of the President's Commission on the Accident at Three Mile Island.* Washing-
           ton, D.C.: President's Commission on the Accident at Three Mile Island.

Pruitt, D., G. Dean, and K. Kressel
    1985   The mediation of social conflict. In D. Pruitt, G. Dean, and K. Kressel, eds., *Journal of Social Issues* 41(2):1-10.
Raiffa, H.
    1968   *Decision Analysis: Introductory Lectures on Choice Under Uncertainty.* Reading, Mass.: Addison-Wesley.
Rayner, S.
    1992   Cultural theory and risk analysis. Pp. 83-115 in S. Krimsky and D. Golding, eds., *Social Theories of Risk.* Westport, Conn.: Praeger.
Reichard, E.G., and J.S. Evans
    1989   Assessing the value of hydrogeological information for risk-based remedial action decisions. *Water Resources Research* 25(7):1451-1460.
Renn, O., and D. Levine
    1991   Credibility and trust in risk communication. Pp. 175-218 in R.E. Kasperson and P.J.M. Stallen, eds., *Communicating Risks to the Public.* Dordrecht, Netherlands: Kluwer Academic Publishers.
Renn, O., T. Webler, and B. Johnson
    1991   Public participation in hazard management: The use of citizen panels in the U.S. *Risk: Issues in Safety and Health* 2(Summer):196-226.
Renn, O., T. Webler, H. Rakel, P. Dienel, and B. Johnson
    1993   Public participation in decision making: A three-step procedure. *Policy Sciences* 26:189-214.
Renn, O., T. Webler, and P. Weidemann
    1995   *Fairness and Competence in Citizen Participation.* Dortrecht, Netherlands: Kluwer Academic Publishers.
Roberson, J., J. Cromwell, S. Krasner, M. McGuire, D. Owen, S. Regli, and R. Summers
    1995   The D/DBP rule: Where did the numbers come from? *Journal of the American Waterworks Association* (October):46-57.
Rodricks, J.V.
    1988   Origins of risk assessment in food safety decision making. *Journal of the American College of Toxicology* 7(4):539-542.
    1992   *Calculated Risks: Understanding the Toxicity and Human Health Risks of Chemicals in our Environment.* Cambridge, U.K.: Cambridge University Press.
Roe, E.M.
    1989   Narrative analysis for the policy analyst: A case study of the 1980-1982 Medfly controversy in California. *Journal of Policy Analysis and Management* 8:251-273.
Rosen, J.
    1990   Much ado about Alar. *Science and Technology* Fall(7)1:85-90
Rosener, J.
    1978   Citizen participation: Can we measure its effectiveness? *Public Administration Review* 38:457-463.
    1981   User oriented evaluation: A new way to view citizen participation. *Journal of Applied Behavioral Science* 17:583-97.
    1982   Making bureaucracy responsive: A study of the impacts of citizen participation and staff recommendations on regualtory decision making. *Public Administration Review* 42:339-345.
Rosner, D., and G. Markowitz
    1985   Public health: Then and now. *American Journal of Public Health* 75(4):344-352.
Ross and Associates
    1991   *Lessons Learned: The Pacific Northwest Hazardous Waste Council's Approach to Regional Coordination and Policy Development.* Seattle, Wash.: Ross and Associates.

Rowe, W.D.
    1977   An Anatomy of Risk. New York: John Wiley.
Rowland, F.S., and M.J. Molina
    1974   Stratospheric sink for chloroflouromethanes: Chlorine atom-catalysed destruc-
            tion of ozone. Nature 249:810-812.
The Royal Society Study Group
    1992   Risk: Analysis, Perception and Management. Chapter 1. London, U.K.: The Royal
            Society.
Rushefsky, M.
    1991   Reducing risk conflict by regulatory negotiation: A preliminary evaluation. Pp.
            109-126 in S. Nagel and M. Mills, eds., Systematic Analysis in Dispute Resolution.
            New York: Greenwood.
Sandman, P.M., N.D. Weinstein, and M.L. Klotz
    1987   Public response to the risk from geological radon. Journal of Communication 37:93-
            108.
Sandman, P.M., N.D. Weinstein, and C.K. Miller
    1994   High risk or low: How location on a "risk ladder" affects perceived risks. Risk
            Analysis 14(1):35.
Schulze, W., G. McClelland, E. Balistreri, R. Boyce, M. Doane, B. Hurd, and R. Simenauer
    1994   An Evaluation of Public Preferences for Superfund Site Cleanup. Unpublished
            manuscript. Department of Agriculture, Resource, and Managerial Economics,
            Center for the Environment, Cornell University.
Sclove, R.E.
    1995   Democracy and Technology. New York: The Guilford Press.
Sewell, W.R., and S.D. Phillips
    1979   Models for the evaluation of public participation programs. Natural Resources
            Journal 19:337-358.
Shapin, S.
    1994   A Social History of Truth. Chicago, Ill.: University of Chicago Press.
Sharlin, H.I.
    1986   EDB: A case study in the communication of health risk. Risk Analysis 6:61-68.
Short, J.F., Jr., and L. Clarke, eds.
    1992   Organizations, Uncertainties and Risk. Boulder, Colo.: Westview.
Shrader-Frechette, K.
    1985   Risk Analysis and Scientific Method. Boston, Mass.: D. Reidel Publishing Company.
    1993a   Burying Uncertainty. Berkeley: University of California Press.
    1993b   Consent and nuclear waste disposal. Public Affairs Quarterly 4:363-377.
    1994   Risk and ethics. Pp. 167-182 in Bo Lindell, ed., Comprehending Radiation Risks: A
            Report to the IAEA. Stockholm, Sweden: Riskkollegiet.
Shrader-Frechette, K., and E.D. McCoy
    1992   Statistics, costs and rationality in ecological inference. Trends in Ecology and Evo-
            lution 7(3):96-99.
Shrivastava, P.
    1987   Bhopal: Anatomy of a Crisis. Cambridge, Mass.: Ballinger Publishing Company.
Siegel, K., and W.C. Gibson
    1988   Barriers to the modificaiton of sexual behavior among heterosexuals at risk for
            acquired immune deficiency syndrome. New York State Journal of Medicine 14:66-
            70.
Simon, H.A.
    1982   Sciences of the Artificial. Second edition. Cambridge, Mass.: Massachusetts Insti-
            tute of Technology Press.

Skaburskis, A.
 1989   Impact attenuation in conflict situations: The price effects of a nuisance land-use. *Environment and Planning A* 21:375-383.
Slovic, P.
 1987   Perception of risk. *Science* 236:280-285.
 1992   Perception of risk: Reflections on the psychometric paradigm. Pp. 117-152 in S. Krimsky and D. Golding, eds., *Social Theories of Risk*. New York: Praeger.
 1993a   Perceived risk, trust, and democracy. *Risk Analysis* 13(6):675-681.
 1993b   Perceptions of environmental hazards: Psychological perspectives. Pp. 223-248 in T. Gärling and R. G. Golledge, eds., *Behavior and environment: Psychological and geographical approaches*. Amsterdam, Netherlands: North-Holland.
Slovic, P., J. Flynn, and M. Layman
 1991   Perceived risk, trust, and the politics of nuclear waste. *Science* 254:1603-1607.
Slovic, P., M. Layman, N. Kraus, J. Flynn, J. Chalmers, and G. Gesell
 1991   Perceived risk, stigma, and potential economic impacts of a high-level nuclear waste repository in Nevada. *Risk Analysis* 11:683-696.
Slovic, P., B. Fischhoff, and S. Lichtenstein
 1979   Rating the risks. *Environment* 21(3):14-20, 36-39.
Slovic, P., S. Lichtenstein, and B. Fischhoff
 1984   Modeling the societal impact of fatal accidents. *Management Science* 30:464-474.
Smith, R.J.
 1986   Chernobyl report surprisingly detailed but avoids painful truths, experts say. *Washington Post* August 27:A25.
Spetzler, C.S., and C.-A. Stael von Holstein
 1975   Probability encoding in decision analysis. *Management Science* 22:340-358.
Starr, C.
 1969   Social benefit versus technological risk. *Science* 165:1232-1238.
Stephenson, M., and G. Pops
 1989   Conflict resolution methods and the policy process. *Public Administration Review* 3:463-473.
Stern, P.C.
 1991   Learning through conflict: A realistic strategy for risk communication. *Policy Sciences* 24:99-119.
Stewart, T.R., R.L. Dennis, and D.W. Ely
 1984   Citizen participation and judgment in policy analysis: A case study of urban air quality. *Policy Sciences* 17:67-87.
Stone, R.
 1994   California report sets standard for comparing risks. *Science* 266:214.
Stulberg, R.
 1981   The theory and practice of mediation: A reply to Professor Susskind. *Vermont Law Review* (6):85-88
Susskind, L., and J. Cruikshank
 1987   *Breaking the Impasse: Concentual Approaches to Resolving Public Dispute*. New York: Basic Books.
Susskind, L., and G. McMahon
 1985   The theory and practice of negotiated rulemaking. *Yale Journal on Regulation* 3:133-165.
Susskind, L., and C. Ozawa
 1985   Mediating public disputes: Obstacles and possibilities. *Journal of Social Issues* (41)2:145-159
Suter, G.
 1993   *Ecological Risk Assessment*. Boca Raton, Fla.: Lewis.

Syme, G.J., and B.S. Sadler
1994   Evaluation of public involvement in water resources planning: A researcher prac-
        titioner dialogue. *Evaluation Review* 18(5):523-542.
Taylor, A.C., J.S. Evans, and T.E. McKone
1993   The value of animal test information in environmental control decisions. *Risk
        Analysis* 13(4):403-412.
Technical Review Committee on the Yucca Mountain Socioeconomic Project
1993   Nuclear Waste at the Millennium: The Human Dimension. Draft Report. State of
        Nevada Nuclear Waste Project Office, Carson City.
Thompson, M., R. Ellis, and A. Wildavsky
1990   *Cultural Theory.* Boulder, Colo.: Westview Press.
Travis, C.
1988   *Carcinogen Risk Assessment.* New York and London: Plenum Press.
Tversky, A., and D. Kahneman
1981   The framing of decisions and the psychology of choice. *Science* 211:453-458.
Tversky, A., and D. J. Kochler
1994   Support theory: A nonextensional representation of subjective probability. *Psy-
        chological Review* 101:547-567.
U.S. Department of Energy
1992   *Draft Final Report of the Secretary of Energy Advisory Board Task Force on Radioactive
        Waste Management.* Washington, D.C.: U.S. Department of Energy.
U.S. Department of Health and Human Services
1993   *Recommendations to Improve Health Risk Communication. A Report on Case Studies in
        Health Risk Communication.* Washington, D.C.: U.S. Department of Health and
        Human Services.
U.S. Environmental Protection Agency
1983   *Community Relations in Superfund: A Handbook.* Washington, D.C.: U.S. Environ-
        mental Protection Agency.
1987   *Unfinished Business: A Comparative Assessment of Environmental Problems.* Office of
        Policy Analysis. Washington, D.C.: U.S. Environmental Protection Agency.
1990   *Reducing Risk: Setting Priorities and Strategies for Environmental Protection.* SAB-
        EC-90-021. Washington, D.C.: U.S. Environmental Protection Agency.
1992a   *Framework for Ecological Risk Assessment.* EPA/630/R-92/001. Washington, D.C.:
        U.S. Environmental Protection Agency.
1992b   *Guidelines for Exposure Assessment.* EPA/600/Z-92/001. Washington, D.C.: U.S.
        Environmental Protection Agency.
1992c   Negotiated Rulemaking on Disinfectants and Disinfection Byproducts. Final draft
        (January 13, 1993). Summary of technical workshop, November 4-5.
1992d   Negotiated Rulemaking on Disinfectants and Disinfection Byproducts. Final draft
        (January 13, 1993). Summary of public meeting, November 23-24.
1992e   Negotiated Rulemaking on Disinfectants and Disinfection Byproducts. Final draft
        (February 24, 1993). Summary of public meeting, December 17-18.
1992f   *Report on the Ecological Risk Assessment Guidelines Strategic Planning Workshop.*
        EPA/630/R-02/002. Washington, D.C.: U.S. Environmental Protection Agency.
1992g   *Peer Review Workshop Report on a Framework for Ecological Risk Assessment.* EPA/
        630-R-92/005. Washington, D.C.: U.S. Environmental Protection Agency.
1992h   *Environmental Equity: Decreasing Risk for all Communities.* Washington, D.C.: U.S.
        Environmental Protection Agency.
1992i   *Safeguarding the Future: Credible Science, Credible Decisions* EPA/600/9-91/050.
        Washington, D.C.: U.S. Environmental Protection Agency.

1993a *Report on the Technical Workshop on WTI Incinerator Risk Issues.* U.S. EPA Risk Assessment Forum. EPA/630/R-94/001. Washington, D.C.: U.S. Environmental Protection Agency.

1993b Negotiated Rulemaking on Disinfectants and Disinfection Byproducts. Final draft (February 24, 1993). Summary of public meeting, January 13-14.

1993c Negotiated Rulemaking on Disinfectants and Disinfection Byproducts. Final draft (June 1, 1993). Summary of public meeting, February 9-10.

1993d Negotiated Rulemaking on Disinfectants and Disinfection Byproducts. Final draft (June 1, 1993). Summary of public meeting, February 24-25.

1993e Negotiated Rulemaking on Disinfectants and Disinfection Byproducts. Final draft (June 1, 1993). Summary of public meeting, March 18-19.

1993f Negotiated Rulemaking on Disinfectants and Disinfection Byproducts. Final draft (July 7, 1993). Summary of public meeting, May 12-13.

1993g Negotiated Rulemaking on Disinfectants and Disinfection Byproducts. Final draft (July 7, 1993). Summary of public meeting, June 22-23.

1993h *Communicating Risk to Senior EPA Policy Makers: A Focus Group Study.* OAQPS Report. Washington, D.C.: U.S. Environmental Protection Agency.

1993i *A Review of Ecological Assessment Case Studies form a Risk Assessment Perspective.* EPA/630/R-92005. Washington, D.C.: U.S. Environmental Protection Agency.

1993j *A Guidebook to Comparing Risks and Setting Environmental Priorities.* EPA/230/B/93/003. Washington, D.C.: U.S. Environmental Protection Agency.

1994a Health Assessment Document for 2,3,7,8—Tetrachlorodibenzo-p-dioxin (TCDD) and Related Compounds. External review draft. 3 volumes. Office of Health and Environmental Assessment and Office of Research and Development, U.S. Environmental Protection Agency.

1994b Explaining Uncertainty in Health Risk Assessment: Effects on Risk Perception and Trust. Draft report, U.S. Environmental Protection Agency.

U.S. General Accounting Office
1995 *Department of Energy National Priorities Needed for Meeting Environmntal Agreements.* (GAO/RCED-95-1). Washington, D.C.: U.S. General Accounting Office.

US Man and the Biosphere Program Human-Dominated Systems Directorate (US MAB)
1994 *Isle au Haut Principles: Ecosystem Management and the Case of South Florida.* Washington, D.C.: US Man and the Biosphere Program.

U.S. Nuclear Regulatory Commission
1975 *Reactor Safety Study.* (USNRC, WASH 1400). Washington, D.C.: U.S. Nuclear Regulatory Commission.

Vari, A., J.L. Mumpower, and P. Reagan-Ciricione
1993 *Low-Level Radioactive Waste Disposal Facility Siting Processes in the United States, Western Europe, and Canada.* Albany: Center for Policy Research, State University of New York.

Vaughan, D.
1990 Autonomy, interdependence, and social control: NASA and the space shuttle Challenger. *Administrative Science Quarterly* 35(2)June:225-257.

Vaughan, E.
1993a Individual and cultural differences in adaptation to environmental risks. *American Psychologist* 48(6):673-680.

1993b Chronic exposure to an environmental hazard: Risk perceptions and self-protective behavior. *Health Psychology* 12:74-85.

1995 The significance of socioeconomic and ethnic diversity for the risk communication process. *Risk Analysis* 15(2):169-180.

Vaughan, E., and M. Seifert
1992 Variability in the framing of risk issues. *Journal of Social Issues* 48(4):119-135.

Viscusi, W.K., and M.J. Moore
   1989   Rates of time preference and valuations of life. *Journal of Public Economics* 38:297-317.

von Winterfeldt, D.
   1992   Expert knowledge and public values in risk management: The role of decision analysis. Pp. 321-342 in S. Krimsky and D. Golding, eds., *Social Theories of Risk*. Westport, Conn.: Praeger.

von Winterfeldt, D., and W. Edwards
   1986'   *Decision Analysis and Behavioral Research*. New York: Cambridge University Press.

Wald, P.
   1985   Negotiation of environmental disputes: A new role for the courts. *Columbia Journal of Environmental Law* 10:1-33.

Wallsten, T.S., and D.V. Budescu
   1983   Encoding subjective probabilities: A psychological and psychometric review. *Management Science* 29:151-172.

Warren, J.M.
   1987   Testimony, Natural Resources Defense Council. House Committee on Energy and Commerce, Subcommittee on Transportation, Tourism, and Hazardous Materials. December 9.

Watson, S.R.
   1981   On risks and acceptability. *Journal of the Society for Radiological Protection* 1(4):21-25.

Weaver, W.
   1948   Science and complexity. *American Scientist* 36:536-544.

Webler, T., H. Kastenholz, and O. Renn
   1995   Public participation in impact assessment: A social learning perspective. *Environmental Impact Review* 15(5):443-464.

Webler, T., D. Levine, H. Rakel, and O. Renn
   1991   A novel approach to reducing uncertainty: The group Delphi. *Technological Forecasting and Social Change* 39:253-263.

Webler, T., and O. Renn
   1995   A brief primer on public participation: Philosophy and practices. Pp. 17-33 in O. Renn, T. Webler, and P. Wiedemann, eds., *Fairness and Competence in Citizen Participation: Evaluating Models for Environmental Discourse*. Dordrecht, Netherlands: Kluwer Academic Publishers.

Whittemore, A.S.
   1983   Facts and values for environmental toxicants. *Risk Analysis* 3:23-33.

Wilson, J.D.
   1994   The use of scientific information in the workplace standard-setting process. In C.M. Smith et al., eds., *Occupational Health Risk Assessment: Directions for the 90s*. Westport, Conn.: Auburn House.

Wilson, R.
   1975   The costs of safety. *New Scientist* 68:274-275.

Wittgenstein, L.
   1989   *Philosophical Investigations*. New York: Macmillan.

Wolpert, R.L.
   1989   Eliciting and combining subjective judgements about uncertainty. *International Journal of Technology Assessment in Health Care* 5(4):537-557.

Wynne, B.
   1980   Technology, risk and participation: The social treatment of uncertainty. Pp. 83-107 in J. Conrad, ed., *Society, Technology and Risk*.

1987   *Risk Management and Hazardous Wastes: Implementation and the Dialectics of Credibility.* Berlin, Germany: Springer.
1989   Sheep farming after Chernobyl. *Environment* 31(2):11-15, 33-39.
1992   Risk and social learning: Reification to engagement. Pp. 275-300 in S. Krimsky and D. Golding, eds., *Social Theories of Risk.* Westport, Conn.: Praeger.
1995   Public understanding of science. Pp. 361-388 in S. Jasanoff et al., eds., *The Handbook of Science and Technology Studies.* Thousand Oaks, Calif.: Sage Publications.

Young, C., G. Williams, and M. Goldberg
1993   Evaluating the Effectiveness of Public Meetings and Workshops: A New Approach for Improving DOE Involvement. Environmental Assessment and Information Sciences Division, Argonne National Laboratory. NTIS DE93HI-019868.

Zeckhauser, R.
1975   Procedures for valuing lives. *Public Policy* 23:427-463.

Zeiss, C., and J. Atwater
1991   Waste disposal facilities and community response: Tracing pathways from facility impacts to community attitude. *Canadian Journal of Civil Engineering* 18(1)83-96.

Index